Evolution and
Human Behavior

A graduate of the University of Wisconsin, ALEXANDER ALLAND, JR. received his doctoral degree in anthropology from Yale University. He has done field work both in the United States and in West Africa and has also taught anthropology at Vassar College, the University of Connecticut, and Hunter College. Dr. Alland presently teaches anthropology at Columbia University in New York City, and is the author of a number of books, including *Human Diversity*, *The Human Imperative*, and *Medical Anthropology*.

Evolution and Human Behavior

AN INTRODUCTION TO
DARWINIAN ANTHROPOLOGY

Second Edition
Revised and Expanded

Alexander Alland, Jr.

Anchor Books
Anchor Press/Doubleday
Garden City, New York
1973

The line illustrations for this book were prepared by the Graphic Arts Division of The American Museum of Natural History.

The original edition of *Evolution and Human Behavior* was published by The Natural History Press in 1967. The Anchor Books revised edition is published by agreement with The Natural History Press.

Anchor Books edition: 1973

For My Parents
and My Colleague
Abraham Rosman

CONTENTS

PREFACE TO THE SECOND EDITION

The occasion of a second edition gives an author the opportunity to reply to his critics. First of all I should like to thank those who pointed out errors and typos in the first edition. I have profited from their careful reading.

For those who expected this book to be a complete introduction to physical anthropology let me say that this was never my intention. Rather, the book is an attempt to bridge a gap between physical and cultural anthropology, between Darwinian theory in biology and works on social evolution and culture change. If anything, this gap has widened in the last several years because of the growing specialization in physical anthropology which has directed it more and more into human biology. The four-field approach which I value has, on the other hand, been held together and strengthened by the work of those physical anthropologists who have concerned themselves specifically with the evolution of culture and those archeologists who have, at least in part, adopted a biological model for their analysis. I have emphasized their work to a much greater degree in this edition.

Half this book is devoted to biological evolution and genetics. The intention here is to inform students of cultural anthropology about the dynamics of the Darwinian model in the context of genetics.

Along with the final chapters it is meant, among other things, as a caution against a too strict adherence to biological determinism. For while my approach is biological and Darwinian I stand firmly with those cultural anthropologists who see our major task as the analysis of culture. For me this can best be done in the framework of Darwin's theory, for man as a species must adapt to his environment just as all other species must adapt to theirs. The task then is to analyze the process of environmental adaptation through cultural behavior and to seek the roots of this behavior in our biological past.

I chose the title "Evolution *and* Human Behavior" rather than "Evolution *of* Human Behavior" specifically to free myself from the responsibility of presenting a full account of human physical and behavioral evolution. There are several good books by physical anthropologists on this topic. I am a cultural anthropologist. I use the biological model of evolution in my work because it helps me to understand human behavior better. I hope it will have the same value for my readers.

While the basic format of this book remains unchanged, it has been corrected, brought up to date, and expanded. The reader of this edition will find a greater emphasis on cultural change and positive feedback. I have also enlarged the sections on behavioral genetics and behavioral evolution because these chapters were excessively thin in the first edition. They are still not meant to present a full account of these fields. Minor corrections and additions have been made in all chapters. Major editions have been made to Chapters VI through X.

I am particularly indebted to Scott Atran who

first suggested a connection between the structural analysis of myth and ecology; to Carole Vance whose work on Irish peasants I have abstracted in Chapter X; and to Ben White whose work on labor and demography has influenced my thinking about population problems.

I should also mention that under the avuncular influence of Robert F. Murphy I have modified my criticism of Julian Steward which was much too harsh in the first edition.

ALEXANDER ALLAND

1973

PREFACE

I have written this book with two purposes in mind. It is structured as an introduction to evolutionary theory and genetics for anthropology students, but it also presents a point of view which I hope will be of some value to my colleagues. This point of view grows out of Darwinian evolutionary theory and represents an extension of it to problems of human behavior including culture. I have attempted to place my view of "cultural evolution" within the context of the biological model and to offer both a criticism and an analysis of previous views of human behavior within the evolutionary framework. The book begins with a section on evolutionary theory and genetics. This is followed by a summary presentation of behavioral genetics and behavioral evolution. The second section is offered as a bridge between classical genetic and evolutionary theory and an analysis of culture. The last section examines social evolutionary theories and presents an extension of the biological model to human behavior. The final chapter offers some potential empirical applications.

In writing this book I have attempted to draw on pertinent data and theory from the fields of biology, ecology, psychology, anthropology, and cybernetics. Cybernetics provides a key to an important aspect of evolution: the development of self-regulating or homeostatic systems. These systems are in effect the outcome of evolution. They must be analyzed from two points of view. First, they must be viewed as structures in which internal organization is geared to efficient operation. Changes in one aspect of internal organization therefore affect other aspects of the system. Second, these systems exist in time and space (their specific environments) and can be fully understood only in terms of the environmental pressures which have affected their development. The environments of these systems consist of both their physical surroundings and other populations of the same and different species with which they interact.

I am indebted to a host of scholars for the ideas presented in this book. My thinking has been particularly stimulated by the works of Theodosius Dobzhansky, Ernst Mayr, George Gaylord Simpson, and W. Ross Ashby. In addition, I owe a debt to Morris Goodman, whose writing on the evolution of primates has been particularly thought-provoking.

Course work with Clellan Ford of Yale University sparked my original interest in the relationships between biology and culture, and conversations with my colleagues at Columbia University, particularly with Andrew P. Vayda, Paul Collins, and Marvin Harris, have helped to sharpen my focus on this subject.

A summer at the Institute of Behavioral Genetics at Berkeley stimulated my interest in the genetic aspects of mammalian behavior.

I also wish to acknowledge the aid of the Department of Anthropology at Columbia and Mrs. Lee Kuhns for correcting and typing the final draft of the manuscript.

I

DARWINIAN EVOLUTION AND GENETICS

All elegant scientific theories have three things in common: they are simple, they explain a great deal, and they can be tested through either observation or experimentation, or both.

Darwin's theory of evolution is one such elegant theory. With only a few basic assumptions it attempts to explain the development and diversification of life, and is thus a major unifying concept in biology. Furthermore, since its development in the middle of the nineteenth century, scientists have accumulated a considerable quantity of evidence in its support.

Darwin's major assumption was that all life is related, and that the number of species occupying the earth has increased through time as a result of continual branching and development from ancestral forms. Evidence for this assumption has been accumulated by paleontologists whose investigations of the fossil record have provided a considerable body of information on the historical development of life on earth.

Taking this first assumption as given, the theory proceeds to account for the process of diversification of living forms. Darwin (and Alfred Russel Wallace, the codiscoverer of the theory) suggested that the variation which exists in nature within and between species is, at least in a metaphorical sense, exploited by variations in the natural environment. Simply stated: Where competition exists for such things as space or food, those organisms most fit to survive and reproduce in a particular environment will reproduce and survive in greater numbers than less well-adapted forms. If the competitive process continues unabated for some time, the less well-adapted forms will be reduced to insignificant numbers or be entirely eliminated from the population. Darwin called this process *natural selection*.

In modern evolutionary theory, comparative reproduction has replaced comparative elimination. Evolutionary success is based on the genetic contribution made by an organism to the next generation. Instead of survival of the fittest or selective mortality, we now speak of *selective fertility*.

The theory of evolution includes two sets of balanced biological phenomena. One set provides relative stability in plant and animal species from generation to generation. Another set contributes some source of variation to plant and animal species. The first are called mechanisms of continuity, and the latter mechanisms of variation. While both sets are necessary for evolution, it is one of the paradoxes of biology that one, the mechanisms of continuity, reflects the perfection of biological systems and that the other, the mechanisms of variation, is nothing more than mistakes or errors in a process of replication. Without these mistakes there could be no evolution, and evolution, therefore, is in a very real sense an accidental process. What this means is that change does not occur in response to need. Nature does not provide species with the inherent ability to adapt to environmental variation. An evolutionary change can occur only if some of the variation already present within the population has a certain value as far as adaptation to new conditions is concerned. Once an adaptive trend has been established, however—that is, once a group of organisms has begun the shift toward a particular sort of adaptive change—this change will tend to continue in the same relative direction so long as there is adequate variation present and so long as the environmental demands remain relatively constant. If for example, the development of an opposable thumb (the thumb opposite the fingers, as in the human hand) provides a group of tree-living animals with a distinct advantage for holding on, as opposed to falling, any variation in the direction of a better grasping hand will have a selective advantage and a trend will be established. Thus a series of random or accidental events can under certain conditions lead to a predictable or nonrandom series of changes. If the environment selects only those forms which are best suited to prevailing conditions the possibilities for change are soon narrowed. As adjustment to a particular environment continues, the chances of widely divergent variations surviving decreases. Only those variations which represent improvement along the line of an established trend are likely to survive. Adaptation represents a goodness of fit between the form of the animal or plant species and the environment in which it must live. Another way of putting this is to say that a group of animals never invades a new environment fully adapted to it since adaptation involves interaction between the environment and certain biological characteristics of the organism.

Species which have recently invaded a particular geographic zone and which therefore have not adapted to it are said to be *generalized* for that zone. As the process of adaptation unfolds, the resulting species become more and more *specialized*. That is, their adaptation becomes more and more efficient for a specific environment. Specialization represents an increase in population efficiency since it reflects successful adaptation. Specialization, however, also leads to restricted variation, since any significant change from a specialized form is highly likely to be less viable than the established type. Thus old, fully established species, well adapted to a specific and rather invariant environment, are likely to change little or not at all. They have "traded" variability for continuity geared to long-standing conditions. Among animals of this type, the cockroach, the ant, the clam, and many simpler organisms are excellent examples. The tenacious adherence to particular adaptive forms which each of these species exhibits is quite amazing.

Some adaptations are extremely narrow, that is, highly specific to particular features of an environment. Thus the aye-aye, a lemur living in Madagascar, eats only one kind of food (a grub living deep in the bark of certain trees) and has a long nail or claw on one finger with which it extracts the food. If the grub were to become extinct, the aye-aye would have no food. On the other hand, the human adaptation (high intelligence, leading to language and culture) plus such generalized mammalian characteristics as warm blood and flexible limbs allow our species to inhabit a wide range of environments. There is therefore no one-to-one relationship between adaptation and specialization even though some specialization is to be expected through time as part of the normal process of accommodation to the environment.

The mechanisms of variation and continuity can be fully understood only with the help of genetics, a science which was unknown in Darwin's time. Unfortunately, while the original theory of evolution is quite simple, genetics is a complicated business. This is true partially because the theory of evolution is very general, and genetics as it applies to real situations is very specific. The contribution of genetics to evolutionary theory lies in its detailed accounting of the sources of variation and continuity which occur within all living organisms. Combined with the study of environment and the fossil record, genetics helps us to understand not only what has occurred in evolution, but how and why it has happened. To understand this, we shall attempt to analyze the relationships that exist between genetics and evolutionary theory.

Figure 1. Dachshund mother with puppies.

First of all, let us consider some examples of continuity and variation as they appear in familiar situations.

It does not take an expert to predict that matings between purebred dogs will produce offspring very much like, but not exactly like, their parents (*Figure 1*). It would indeed be a surprise if a properly mated Great Dane were to labor mightily and bring forth a chihuahua, or a fox terrier, for that matter. On the other hand, crosses between two different varieties of dog will produce offspring of greater variation and two mongrels of unknown origin will produce quite an unpredictable array of puppies (*Figure 2*). In all these cases, however, there is a considerable degree of continuity as well. All dogs produce dogs and not cats, and Great Danes produce dogs which are also Great Danes. Breeds, or subspecies, can be interbred to produce mongrels, but species generally cannot be crossed to produce intermediate varieties. In those few cases in which this rule is broken (tigers can breed with lions and horses with donkeys) the offspring are infertile. Thus the species is a closed unit. Members of a species can interbreed successfully with each other, but they cannot produce fertile offspring if they breed with the members of some other species.

The source of much of the observed continuity is *hereditary;* that is, the reproductive process involves the transfer of invariant *genetic* material from parent to offspring. Each breed of dog, however, has thousands of units of genetic material which control such traits as size, shape, coat color, length of tail, etc. In some cases a single unit controls a trait; in other cases many units combine to produce a trait.

Figure 2. Top: *Boxer parents with puppies. The parents have had their tails and ears cut, but acquired characteristics are not inherited and the newborn puppies have the long ears and tails.* Bottom: *Mongrel parents and their mongrel offspring.*

Within any species some of these genetic units are held in common by all members of the species, while others are specific to specific breeds. The distinctive attributes of dogs are the result of genetic units distributed throughout the species. The distinctiveness of Great Danes is the result of units distributed within a single breed. Still other genetic units are restricted further in their distribution and account for individual differences within a breed. These units or *genes* are generally passed on unchanged from parent to offspring, but an offspring receives only one half of its genes from each parent. When two purebred dogs of the same strain mate, much of the genetic material passed on to the offspring from each parent is similar and little variation results. When mongrels mate, the situation is somewhat different. Genes are passed on unchanged to the offspring, but these genes themselves represent a wider variety of traits because some represent one breed, some another, some still another. These units combine at random in the puppies, and greater variation results. It is important to emphasize that the genetic units themselves do not change from one generation to the next. It is only the combination of these units which is different. The variation which is observed is due to new combinations of the invariant genetic units.

Unfortunately, a further complication emerges when we examine variation more closely. A good deal of difference can be produced among littermates by varying certain nonhereditary conditions, such as the amount and kind of food given to different dogs, or the amount of exercise allowed to each. When we vary these conditions we are changing the environment. Some animals may be brought up in one type of environment and some in another. Any variation which results from this type of manipulation is the result of environmental factors. It is sometimes difficult to sort out which differences are due to environment and which are due to heredity. One way of separating these two factors is based on the observation that environmental differences (variations acquired during the lifetime of the animal) are not passed down to the next generation. This is another way of saying that acquired characteristics are not inherited. Breeding experiments can reveal which traits are hereditary. Thus we know that boxer dogs have been deprived of most of their tails and a good part of their ears for as long as boxer fanciers have enjoyed docked ears and tails, but that boxers are always born with large ears and long tails. Generations of surgery have not altered the heredity of boxers (*Figure 2*).

Any differences produced by environmental variation fall within definite limits, for while the environment can strongly influence the development of an individual animal, it can never transform it into something which lies beyond the boundaries of its own heredity.

Development, then, is determined by a combination of relatively constant factors which are part of the organism's own potential, and by external conditions which make up the life experience of the organism. Thus every creature is a product of its own particular environmental history and a part of the genetic history of its ancestors.

Three examples of the interaction between heredity and environment can serve to clarify what we mean here.

Under normal conditions, one of the distinguishing features of Siamese cats is a type of pigmentation in which the animal's body is generally light in color, with progessive darkening toward the tips of the feet, ears, and tail. This pattern comes chiefly in two varieties known as "seal point" and "blue point." When cats that look Siamese are mated they will always produce kittens of the same general color pattern. The chemical process which underlies the particular pigmentation of these animals is highly sensitive to temperature. Siamese cats are darker at the extremities because it is these areas of the body which are coolest. Almost white cats could be produced by keeping animals in an extremely warm environment, and these could then be turned into nearly black cats by lowering their body temperature sufficiently to overcome the normal suppression of pigment. Few people, if any, keep their Siamese cats in either the oven or the refrigerator, however, and I doubt if anyone has seen an un-Siamese-looking Siamese cat.

A less trivial example is that of diabetes in man. This disease is related to a breakdown in the functioning of the pancreas, an organ which, among other things, operates in carbohydrate metabolism. Few people are born diabetic, and yet diabetes is recognized as a hereditary disease. What is actually inherited is a weakness in pancreatic structure which may lead to a breakdown, particularly if an affected individual overindulges a taste for carbohydrates. The disease is the result of environmental stress on a potentially inoperant organ. Furthermore, since the discovery of insulin, the hormone involved in carbohydrate metabolism, a patient's internal environment can be altered artificially through inoculation of this hormone to restore correct body processes and suppress the effects of the disease. All "curable" hereditary diseases are treated by altering the internal environment of the affected individual in a way that will restore normal function and thus suppress the hereditary malfunction.

Some animals, particularly the chameleon, have a peculiar hereditary mechanism which is keyed to environmental stimulation. These animals are much more variable than Siamese cats, and chameleon watchers often catch them in the act of changing their color markedly from one moment to the next. For a long time it was believed that these pigment changes were a protective response to background color, and that, in effect, the animal had a built-in device for instant camouflage. Actually these changes are due to other types of environmental stimulation, such as light intensity and temperature changes, as well as excitement caused by such external conditions as fright. These environmental variables trigger a hormonal response in the organism, and a color change results. Exactly what these reactions have to do with survival is still a debated question, but the fact remains that a chameleon's color at any particular time is a function of inherent body chemistry plus certain external conditions.

Siamese cats and chameleons each in their own way demonstrate that the interaction pattern between genetic and environmental variables may be highly plastic. This is not always the case, however, and the fact that organisms may be irreversibly shaped by environmental factors must not be overlooked. For example an animal which has been stunted during its normal period of growth will never recover its hereditary potential, and "normal" growth itself is just as much the result of early external influences: in this case, those which maximize the growth potential of the organism. The result of early environment-heredity interaction can set the pattern for the entire life-span of the individual. Ample evidence for this is provided by a series of metabolic diseases in man which lead to permanent mental deficiency unless they are treated early. Among these are thyroid hypofunction, the effect of which is known as cretinism; and phenylketonuria, which may lead to severe mental impairment. Hypothyroid conditions can be treated with injections of thyroxin, a hormone which is normally produced by the thyroid gland. Phenylketonuria can be treated by decreasing the protein level of the diet to a point where metabolic poisons produced by the hereditary disorder are reduced to a minimal level. In both diseases any brain damage which results in untreated patients is permanent, and no amount of medication can repair the malfunction once it has been allowed to develop.

These examples could be multiplied indefinitely. It is for these reasons that biologists now agree that the argument over the primacy of environment or heredity in the development of organisms is a dead issue. It is now generally accepted that the function and form of organisms can be understood only as the result of a highly complicated process of interaction. The so-called "nature-nurture controversy" is dead except in the minds of a few unsophisticated individuals.

Genotype and Phenotype

For purposes of analysis, however, the genetic background must be separated from environmental effects. This can be done experimentally by raising organisms in the same environment (i.e. holding environment constant) and noting the genetic variation; or by producing a pure strain of organisms with practically identical heredity, manipulating the environment, and noting its effect on variation. Purely genetic effects can also be analyzed by observing hereditary continuity from generation to generation in controlled breeding experiments. Technically the genetic background of the organism, which can include a range of unexpressed traits, is called the *genotype*. The genotype represents the hereditary potential of an organism. Heredity (the genotype) acting in combination with a particular environment produces the *phenotype* or product of interaction. Another way of expressing this is to say that the phenotype is the result of a particular heredity acting on a particular environmental background. Any variation we observe among the members of a related group of organisms living under natural conditions must be phenotypic variation, because it will be the result of different environmental pressures and different genetic histories. Phenotypic variation in a population is the sum of genotypic variation inherent in the combined heredity of the group plus that part of environmental variation which affects the phenotype. This can be expressed in the formula $P_v = E_v + G_v$, in which $P_v =$ phenotypic variation, $E_v =$ environmental variation, and $G_v =$ genotypic variation.

From this formula one might surmise that a high degree of phenotypic variation in any given population will occur as the result of high genetic variation, high environmental variation, or a combination of both.

Indeed this is often the case, but it must be noted that the genetic background itself can determine how much and what kind of environmental variation the phenotype can absorb. Some species are highly susceptible to environmental differences; others can remain stable under a wide range of conditions. Stability can result from two different genetic processes. In some species even small environmental variations are not well tolerated. If a population of such a species is exposed to changed conditions, it will die out. Many microorganisms (bacteria, for example) are so sensitive to such conditions as level of acidity or temperature that they will tolerate only minute differences. Such species will be found only where environmental variation is low, and the phenotypic variation will always be low as well. On the other hand, the genotypic background of other species may be geared to absorb a wide range of environmental variation without developing changes in the phenotype. Man, relatively speaking, is this type of organism.

When we examine the possibilities of stability and variation in both phenotypes and genotypes we find three basic types of population: (1) those which cannot tolerate change in the environment and which are therefore found only under a strictly limited range of conditions; (2) those which remain phenotypically similar in a wide range of environments; and (3) those in which there is a high degree of phenotypic variation based on high genotypic variation and/or a high degree of environmental variation. These three conditions represent different kinds of adaptation, and are, therefore, by-products of evolution themselves. In the first case the species is extremely well adapted to specific conditions. Organisms of this type which are intolerant of external change usually do extremely well under conditions which are optimal for them. They are highly specialized. Flexibility is sacrificed for goodness of fit. In the second case peak efficiency has been "traded" for the ability to survive under a range of conditions. Such organisms are less well adapted to a single set of conditions than the first group, but they do not face extinction if minor changes in the environment occur. (A drastic enough change in the environment will, of course, kill any living thing.) In the third case there are two possibilities. (A) Subgroups of a species vary widely in response to local conditions. Genotypic flexibility is reflected not in phenotypic stability (as in group 2) but in phenotypic variation. This type of adaptation, although very common, generally develops in situations where subgroups are somewhat separated spatially so that different breeding groups can develop. Each group represents an adaptation to local conditions. Some of the variation between human populations is probably due to such local geographic adaptation. (B) The other possibility in the third kind of adaptation occurs when relatively ho-

mogeneous genotypes react strongly to rather small
environmental pressures. There are, for example,
many species of flowers which will exhibit one color
under one set of conditions and another color under
other conditions. Such differences in phenotype are
the result of environmental variation only, acting on
a highly sensitive genotype (*Figure 3*).

This latter case provides an example of the phenomenon known as *phenocopying*. A *phenocopy* is

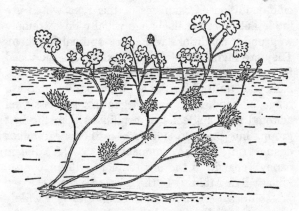

Figure 3. The water crowfoot (Ranunculus aquatilis),
*an example of the plasticity possible in a phenotype: the
leaves above water develop a different appearance from
those below the water level. (After Weaver and Clements)*

an environmentally produced analogue of some genetic property. Thus, to give another example, in
certain bacteria two closely related species grown
on the same media can be differentiated by the color
of their respective colonies, one being white and the
other yellow. As long as they remain on the media

the phenotypes yellow and white are clearly distinguishable. If, however, the white species is placed in another environment, i.e. a different media, it will produce a yellow phenotype. If the yellow phenocopy is returned to the original media it will revert to its original white form. This reversion proves that the phenocopy was produced by environmental variation and not by genetic change. The occurrence of phenocopies demonstrates clearly that multiple routes to the same trait can exist and that particular phenotypes must be seen as outcome products of environment-heredity interactions.

It follows from evolutionary theory that environmental pressures act on all populations to select out the best-adapted phenotypes. If genotypic variation is submerged in a single phenotype, the variation itself can be preserved in the adaptive process. If, on the other hand, most genetic variation is expressed in variant phenotypes, then genotypic variation will be lowered by selection at the same rate as phenotypic variation is lowered. It must be made clear that in both situations the environment acts directly only on the phenotype, on what is expressed. Any effect on the genotype is indirect. As we shall see later, this is of great importance in evolution, since the genotype can represent a reservoir of untapped variation which can be crucial to survival under critical conditions.

What I have been talking about in these last paragraphs is natural selection, for natural selection represents the effect of the environment on the phenotypes of a specific set of organisms, a population. Within the range of variation those organisms which are phenotypically best suited to the environment will have a selective advantage over the others. This does not mean that every single member of a population which has a selective advantage will survive, or that every member with a selective disadvantage will perish. No biologist would care to predict which individual organisms in a population will live to reproduce, since it is impossible to predict all the events which will occur in the life of an organism. Chance events (in effect, accidents) happen even to the best-adapted creatures. A few of the least well adapted may be effectively protected by accident as well. What the biologist attempts to predict is the success value of a particular phenotype occurring in a group of organisms with variant phenotypes. This means that if two or more types exist in the same environment, it may be possible to identify which type will survive in greater numbers than competing types.

Thus, while all organisms as individuals are the products of evolutionary development, evolutionary theory makes predictions about populations, and not about individuals. Statements made about natural selection are called statistical predictions because they involve probability and are never absolute. They can never give an investigator the assurance that a particular individual will survive. Such predictions represent a percentage figure based on the expected selection value for a specific trait or phenotype. Of course there are situations in which a trait is 100 per cent fatal at or near birth. In these special cases it can be said with certainty that any organism born with such a trait will not live to reproduce itself. Its reproductive potential is said to be zero.

This fact does not defy the rule that one cannot predict which organisms will live to reproduce, however! It should be obvious from the above that selective advantage is always a comparative figure. There is no such thing as an absolute selective advantage because selection is measured by comparing the reproductive potential of one group against the reproductive potential of another group or groups. As we shall see later, statistics are a very important part of evolutionary theory, particularly when genetic analysis is involved.

Statistical statements about survival value are always made in reference to a specific environmental background. The over-all picture of evolutionary development which has been so well analyzed by paleontologists and geneticists is made up of small pieces each representing a population, changing through time, each contributing in a small way to the over-all process of development. This is an extremely slow process, and it is therefore difficult to observe it in action. Genetic changes are minute, and a long time is usually required before their additive effects can be demonstrated. Some small but important adaptive changes have been documented, however, particularly in species which reproduce very rapidly and which therefore can be studied through several generations by a single investigator in a reasonable amount of time. Many of these changes have been observed in the laboratory, but occasionally the attention of scientists has been drawn toward naturally occurring phenomena which can only be explained as examples of natural selection. Three famous cases will, I think, suffice as examples. The first involves a dangerous and common parasite of man, *Micrococcus pyogenes* variation *aureus*. These bacteria, also known as *Staphylococcus aureus*, usually live on the human skin and interfere little with the health of their host. Occasionally these bacteria do invade the body. When this happens they are able to cause not one but several diseases ranging from mild boils to severe blood poisoning. The discovery of penicillin was a temporary disaster for these parasites, and for some time they were readily killed by rather low doses of this antibiotic. Unfortunately for man, however, a variant form of this organism existed which was resistant to penicillin. The frequency of the resistant strain was low in the preantibiotic past, and it is therefore quite probable that such organisms have a lower selective value than non-resistant forms. The discovery of penicillin brought about a drastic change in the microorganisms' environment and this shift changed the selective advantage of the existing strains. The result was the development of the so-called "hospital Staph." The rapid spread of this strain is evidence enough of its high selective advantage in an environment which includes antibiotics.

More complicated organisms than bacteria have also provided rather unwelcome examples of adaptation to new environmental conditions. The Australian rabbit represents a particularly interesting case.

A rather insignificant number of rabbits came to Australia with the early settlers from Europe. In retrospect this was a significant mistake, for the descendants of these rabbits soon came to outnumber the descendants of the human immigrants. Unencumbered by the predations of natural enemies, their population explosion reached great proportions. In the new environment rabbits exhibited the full growth potential of a rapidly reproducing species, and rabbits have been the major plague of Australia ever since. In fact, Australia is the only nation that has intentionally divided itself by building a fence almost the entire length of the country. A tremendous percentage of the national budget is spent each year to keep this barrier in good repair, and a group of professional fence watchers lead a lonely life in the wastelands of central Australia. All this to deter the extension of the rabbit population into new territory where they would rapidly destroy the productivity of the land. Several years ago a more economical way of restoring the balance between man and rabbit was discovered—the disease myxomatosis, a deadly scourge of European rabbits. Introduced into the Australian rabbit population, myxomatosis soon destroyed the majority of animals—but not all. Unfortunately, a genetic variation existed which brought resistance to the disease. Resistant rabbits, at first small in number, soon came to replace their less fortunate susceptible relatives, and the long fence is once again the major defense against the wild-rabbit population (*Figure 4*).

Leukemia is now treated with several drugs which destroy most but not all of the malignant cells. Like the Staph bacteria and Australian rabbits, there are small numbers of resistant forms in the population of leukemia cells. When drug therapy is employed, all but resistant cells are destroyed, but these reproduce rapidly and the

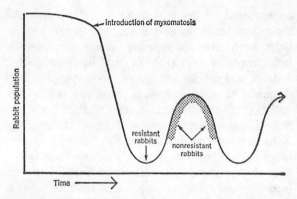

Figure 4. The effect of myxomatosis virus on the Australian rabbit population.

symptoms of this disease eventually reappear. With leukemia as with bacterial diseases, the reproductive rate of the disease agent is extremely rapid. Several generations can develop in only a few days, and rare variant forms soon replace more common members of the population.

Speciation

So far all the examples of variation presented in this chapter have been contained within a definite range of continuity, that of the species. I have discussed variations based on heredity and environment and the process of natural selection through which these variations are exploited, but I have also stressed that dogs remain dogs; cats, cats; and rabbits, rabbits. There appears to be some sort of boundary to genetic variation, a boundary maintained by the integrity of the species. This is due to the fact that interbreeding between species does not normally produce fertile offspring, hence genes cannot be spread through intermixture from one species to another. If this is true, then an important question arises. How can the Darwinian theory, even with the help of genetics, explain the changes that must have occurred across the dividing line between species? How can the origin of species be explained from the principles discussed thus far? This is one of the most important questions in evolutionary theory, and it is the one most frequently asked by Darwin's critics, most of whom will admit that adaptation does indeed occur. In fact, some of these critics have exploited natural genetic variation in breeding better egg-laying chickens or more obedient hunting dogs. Until *speciation,* the transformation from one species to another, is explained and until experimental evidence for its occurrence has been presented, they refuse to admit any link between what they know to be true about animal breeding and Darwin's theory. Essentially they conceive of differences within species (for example, breed differences) as differences of degree, and differences between species as differences of kind. By definition, species are reproductively isolated groups or closed units while breeds of subspecies are open units among which breeding can occur. Reproductive isolation may be due to genetic or behavioral factors, but our main concern here will be with genetic factors. A full discussion of speciation can be found in Mayr's *Animal Species and Evolution.*

The demand for an explanation of speciation and

evidence to support it is a just one. If evolutionary theory could not account for the transformation of species it would be useless, since this is its major task. Fortunately there are two solutions to the problem, two separate lines of evidence which demonstrate that speciation has occurred. One is historical and comes from the fossil record; the other is genetic and is found in the study of contemporaneous populations.

Fossils and Speciation

In the first place, the fossil record allows us to trace the sequential transformation of particular species. In the second place, it provides comparative material illustrative of branching evolution from parental forms. Sequential transformation and branching are both best explained as evolutionary adaptation to environment. In the first case (known as *anagenesis*) a single type develops greater and greater genetic specialization through time in areas where the environment has remained relatively constant (*Figure 5*). This is the progressive development of goodness of fit. In the second case (known as *cladogenesis*) related isolated populations differentiate from one another as a response to differential selection pressures which arise in different microenvironments (*Figure 6*). When these groups become reproductively isolated—that is, when they can no longer interbreed—the genetic differences between them have become great enough for them to fulfill the accepted definition of species.

There are countless examples of anagenesis and

cladogenesis in both the plant and animal kingdoms. The most famous example of a developmental sequence in the fossil record is that provided by the series that led from a rather insignificant four-toed quadruped to the modern one-toed horse. The fossils in this series lend great support to Darwin's theory. In fact, they alone are credited with tipping the balance in Darwin's favor among some nineteenth-century American biologists who were originally skeptical of the theory.

Branching from common ancestors (cladogenesis), as well as progressive speciation (anagenesis), is documented quite well also by the record of the primate order, which includes man, monkeys, and apes. The common ancestors of these forms adapted more than seventy-five million years ago to life in the trees, in which stereoscopic vision and a pair of good grasping hands are advantageous. These are both major features of all but the oldest and most primitive members of the primate order. As the primates developed, they spread into different environmental zones. Some returned to a ground-dwelling existence and modified their earlier adaptations. Among these were the baboon and man, two species which have accommodated in different ways to terrestrial life. The human sequence shows increasing brain size, fully erect posture, and the loss of hands on the lower extremities (*Figures 7 and 8*).

In the last twenty years an impressively large group of fossils has been accumulated which help to fill in the details of man's evolution from earlier primate forms. This evidence would have been gratifying to Darwin. In his book *The Descent of*

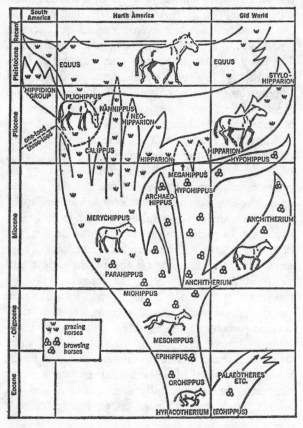

Figure 5. Anagenic development of the horse family. (After Simpson)

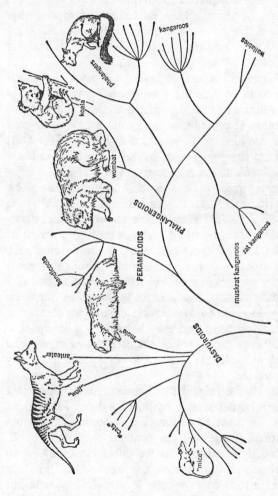

Figure 6. Adaptive radiation among the recent Australian marsupials.

Man Darwin placed the human species among the other members of the animal kingdom and suggested that human evolution has followed the same rules set down in his earlier work, *The Origin of Species*.

The fossil record is useful not only because it allows us to examine a substantial part of a family tree through the course of its development, but also because it provides a good time sequence as well as some idea of the spatial distribution of forms. It is from the evidence of paleontology that we know that some of the ancestral relatives of modern elephants lived in the arctic, including the New World, and that the camel was once native to North America. Time sequences and spatial distributions are important because they give scientists some idea of the time required for development, the order of development, and the range of geography and environment in which a particular evolutionary change has occurred. Geologists have expended much effort to create devices for measuring geological time, and they have been quite successful. These geological clocks help the paleontologist to date his fossil finds and to place them in orderly sequence.

In all the related fossil series it is assumed that genetic differences provided the basis for change. But we must remember that genetic changes are random events and are generally of small quantitative significance. It is for this reason that evolution requires a tremendous amount of time for change to become readily apparent. The earth has been in existence, however, for over four billion years, and life for over half that time. Considering that man is a new species (something over one and a half million years old) two billion years are certainly long enough for all of evolution to have taken place.

There is another lesson in the fossil record—one which supports the contention that evolution based on genetic change is an accidental random process. This is the fact that in the time since the origin of life a tremendous number of plants and animals have become extinct. Species have often followed the "wrong path" in the maze of survival and arrived at a dead end. Environmental changes which were not matched by the independent process of genetic change led to the extinction of many species. The argument that evolution is a directed process, guided by some greater wisdom, is contradicted by the existence of so many "mistakes" in the history of plant and animal development. Things look perfect only to those who see what remains after a sometimes tremendously wasteful process of elimination has continued through an almost unbelievable passage of time. This can be said to be the best of all possible biological worlds only in the sense that what is is possible—as the end product of a long interaction between internal genetic factors and external environmental conditions. Only a goodness of fit between these factors can yield survival and survival is always provisional, dependent upon changing conditions. A simple shift in the environment such as the introduction of a new or foreign species can so upset the balance and "good sense" of nature as to destroy a good part of the natural environment. This is just

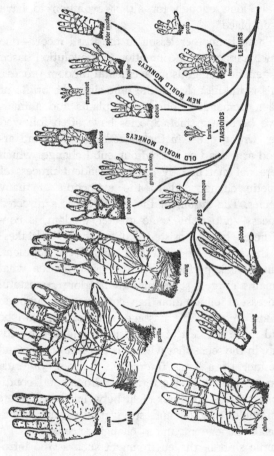

Figure 7. The evolution of the hands of primates. (After Gregory)

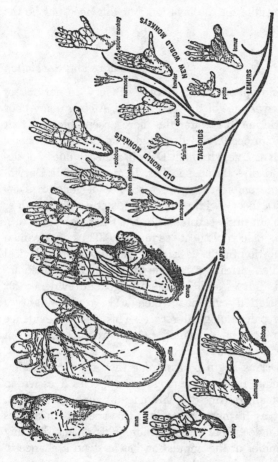

Figure 8. *The evolution of the feet of primates.* (After Gregory)

about what happened in Australia with the introduction of rabbits. In this example, man was responsible for the invasion, but this need not be the case.

Genetics and Speciation

Some opponents of Darwin's theory do not accept the evidence of the fossil record. Among many poor criticisms they offer the more acceptable one that it is impossible to derive genetic data from bones. This is true, although Darwin's theory does not depend directly on genetics. Remember that Darwin wrote before there was a science of genetics. Darwin's theory with or without genetic evidence does explain quite satisfactorily what has been found in the paleontological sequences. Thus Darwinian evolution fulfills its scientific task by offering logical and empirically justifiable explanations for phenomena. But in addition there is genetic evidence for speciation—evidence which adds great weight to the paleontological record. This genetic evidence can best be explained by the examination of an abstract case.

Let us begin by assuming that a sexually reproducing, relatively homogeneous species lives within a small, restricted environmental zone (*Figure 9*). Let us further assume that this species constitutes a single breeding population. This means that any member of the population has as good a chance to mate with any other mature member of that population of the opposite sex. Such a situation would ensure a wide and random distribution of genes in

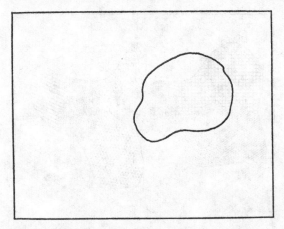

Figure 9. Small panmictic population occupying restricted environmental zone.

the population at large. Random mating of this type is known as *panmixia*. When it occurs the population is said to share a common *gene pool*. Any variation which exists in such a population will be distributed fairly evenly within the confines of the total group. Now if this is a particularly successful species and it spreads out geographically, it is likely that subpopulations will develop as units of the larger group. If the distance between these groups widens, they will eventually constitute separate breeding groups. This is largely a mechanical situation. Animals which are closer together are more likely to breed than animals which are far apart. If any barriers develop between units, then these units will become at least partially isolated (*Figure 10*). In such situations new genetic variations will be unequally distributed in the species at large.

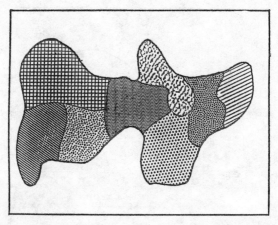

Figure 10. Spread of species into semi-isolated breeding populations.

That is, each subgroup will begin to develop its own gene pool different in some respects from all other gene pools. If the geographic space in which the species is distributed is uneven—that is, if there are environmental variations to which the species is sensitive—then different selection pressures will further differentiate the gene pools of the subgroups. As long as some interbreeding continues to occur between these subpopulations there will be no differentiation beyond the level of the species. Each individual unit will constitute a separate breed or strain of the species. If, however, for some reason some of the populations become totally isolated, they will continue to change to the point where genetic differences will be great enough to produce new species. As I have pointed out above, the reason most species cannot interbreed successfully is because they are rather radically distinct units, units between which there is a sizable amount of genetic difference. As long as gene flow continues, enough genetic similarity will be preserved between strains to stop the process of speciation from reaching finality. Under natural conditions subpopulations tend to become separated through such centrifugal processes as differential genetic variation, differential selection pressures, and semi-isolation, but they are also frequently drawn together by the centripetal process of gene flow. Speciation occurs when the centripetal forces are interrupted.

Now let us imagine our species to be widely distributed in space so that several subpopulations are scattered over a large area. Assume that we are interested in gene flow and differentiation as they occur in a lineal direction from one end of the geographic space occupied by the species to the other. In such a situation it would make sense to assume that gene flow occurs as genetic transfer from one subpopulation to the next. If we label our adjacent groups A, B, C, D, E, F, G, H, then it should be obvious that some interbreeding will occur between groups A and B, between B and C, between C and D, etc. Genes from A get to population H not directly but via a chain of breeding between adjacent populations. If each group is a partial genetic isolate—that is, if most of its breeding takes place within the group—one would expect a diminution of genes from A as they pass through B, C, D, E, F, and G to H (*Figure 11*). A particular genetic trait with a high frequency in A would have lower frequency in B, still lower in C, and so on. If, in addi-

tion, environmental pressures were different on
each population, there might well be a further diminution of the trait as one moved from an area of
high selective advantage to an area of low or negative selective advantage. Such a distribution is
known as a *cline*. Clines are quite common and
many have been analyzed for widely distributed
species.

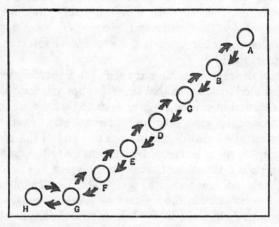

*Figure 11. Clinal distribution of a trait in populations of
a single species widely distributed in space. Arrows indicate gene flow.*

Clinal distributions also tell us something about
populations of a species, for while clines refer to the
distribution of a single trait in a species they often
reflect a scale of difference between populations.
When the clinal difference between A and H in our
example becomes very great, it may well be impossible for individuals from A to produce fertile off-

spring if they are mated to individuals from H. This would be true if many other gene differences appeared along the clinal distribution between A and H. This is often the case.

This presents a problem. Are A and H members of different species, or do they represent variant strains of the same species? The answer depends upon which evidence one might wish to employ in the genetic definition of species. A and H are members of a common gene pool, but genes can only be exchanged between them through intermediate populations genetically more similar to one or the other. If in the type of clinal distribution discussed here, one population, say G, is wiped out, then the gene flow between A and H will cease (*Figure 12*). Then the genetic difference between A and H will

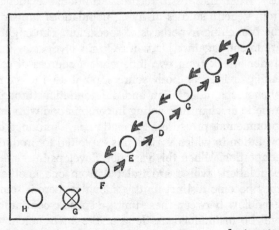

Figure 12. Interruption of gene flow producing two separate species.

widen at a greatly accelerated rate. Furthermore, with the disappearance of population G, no connection between A and H will exist to link them naturally as former members of a common gene pool.

Semi-isolation of populations defined along clinal distributions has frequently been observed in broadly distributed species. Such a phenomenon illustrates that what appear at a single point in time to be differences in kind may in historical perspective be reduced to differences of degree. What is required is evidence for the intermediate steps which must exist or have existed between all related species.

As far as speciation is concerned then, we have evidence of developmental adaptation (anagenesis) and evidence of branching adaptation (cladogenesis) from the fossil record. Additional evidence from genetic studies of living populations supports the branching hypothesis of speciation. The mighty arguments against Darwin's *Origin of Species* fall under attack from two independent sources of scientific evidence. Both sources point to the same phenomena, adaptation and differentiation through genetic mechanisms, acting in combination with environmental pressures. Speciation is a product of evolution in which variation is exploited to produce adaptation. When this variation develops in related populations which are isolated from one another, they become distinct units. As long as there is some gene flow between these units, a specific continuity will be maintained. As long as the environment is stable, this continuity will be reinforced by the fact that only certain genetic combinations will prove adaptive under specific conditions. When the environment varies beyond certain limits, the phenotypic variation inherent to some degree in all populations will either readjust to the new conditions or die out. Extinction results when none of the inherent genetic variation fits the new environmental demands.

Summary

Darwin's theory of evolution assumes that all life is related and that differentiation has occurred as the result of adaptation to environmental demands. The fossil record proves that there has been an increase in the total number of species through time, and at least suggests that forms are connected since it provides some of the missing links between divergent species. Examination and experimental manipulation of living species show that there are variant types in all species and that this variation is based on the hereditary potential of all organisms and on the effects of environment on this potential. Genetics has demonstrated the close relationship between environment and heredity. Selection experiments and observation of natural processes demonstrate that natural selection does indeed occur. Finally, observations of natural species distributed in space plus controlled breeding experiments show that speciation is a process of gradual separation and genetic differentiation. Darwin's theory is supported by paleontology and genetics, two quite different fields of scientific endeavor. Furthermore, the two sets of mechanisms to be demonstrated in this book, variation and continuity, appear to be the dynamic forces behind evolution. Variation provides the material for new adaptations; continuity protects the adjustment which develops between hereditary potential and environmental conditions.

II

MENDELIAN GENETICS

Probably since the earliest domestication of plants and animals man has used certain genetic principles without fully understanding them. Species were improved by breeding selectively only those animals which carried desirable traits and by disposing of those animals which did not. Thus, even today the life expectancy of a good egg-laying chicken is considerably higher than her less bountiful sisters.

To understand how selective breeding actually works, one must understand how and in what frequency specific characteristics pass from parent to offspring. Traits do not always appear to pass in orderly fashion. Two parent animals, each with the same trait, might transfer it to all their progeny while two other parents, again with the same trait, might transfer it with a frequency of only 50 per cent. On the other hand, an entire generation of siblings might differ considerably from either parent. Until the reasons for this variation were understood it was difficult to predict what the outcome of any specific cross might actually be.

A set of simple but elegantly controlled experiments conducted by the Abbé Mendel in the 1860s solved this basic problem and opened up the entire field of genetics. Unfortunately Mendel's work, although published in 1866, was ignored by the scientific community until 1900. The development of genetics, which progressed rapidly after this date, was tremendously important to evolutionists, for it provided a central concept of evolutionary theory: an explanation of the mechanisms of variation and continuity.

The science of genetics which Mendel founded is tremendously powerful, for its theories can be tested and verified in the laboratory through the replication of experiments under controlled conditions. Thus, with the development of genetics a laboratory science was added to the natural-historical approach of the early evolutionists.

Mendel's experiments were carried out on the common pea plant. In his search for order in the transmission of genetic material, Mendel chose pea plants displaying seven classes of easily observable traits, each with two variants. Among these were: height (tall vs. short); seed color (green vs. yellow); and seed texture (smooth vs. wrinkled). Mendel was careful to choose plants of pure strain, which when crossed with like plants bred true consistently, i.e. yellow plants produced 100 per cent yellow offspring, short plants produced 100 per cent short offspring, etc. Plants with various combinations of these three classes of traits were then crossed (tall, green, wrinkled, with short, yellow, smooth, for example) to produce a filial or F_1 generation. Members of this generation were again crossed to produce still another filial or F_2 generation. After the completion of each cross, the frequencies of the resulting traits were carefully recorded and analyzed statistically.

Segregation

Mendel's first concern was with traits of a single class. In these experiments he found that one trait in each class would fail to appear in the F_1 generation. A yellow crossed with a green, for example, would yield an F_1 generation with 100 per cent yellow-seeded plants. Furthermore, when F_1 plants were crossed with each other, the green color would reappear in 25 per cent of the F_2 offspring; the other 75 per cent would, of course, be yellow (*Figure 13*). It appeared that the yellow was able to mask the green, at least in the F_1 generation. Mendel accounted for this fact by suggesting that

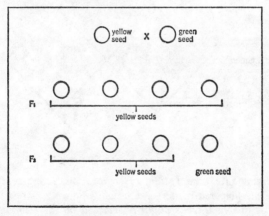

Figure 13. A cross of a yellow with a green seed; the yellow trait is dominant here and masks the green color in the F_1 generation. In the F_2 generation the green color reappears in 25 per cent of the plants because no yellow component is present.

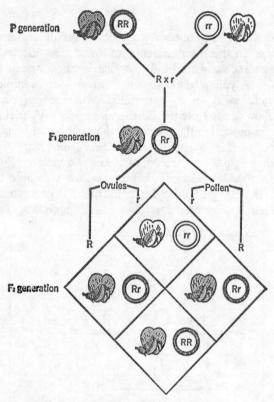

Figure 14. Mendel's law of segregation. A cross of a purple-flowered strain and a white-flowered strain of peas in which traits might be masked by the dominant component (purple) but never combine to form an intermediate form.

some traits were *dominant* over others. The reappearance of the green color in the F₂ occurred because in 25 per cent of these plants no yellow component was present. The green component had "segregated out." Traits which could be masked in this way were said to be *recessive*. Mendel also found that in no case did a cross between green and yellow, tall and short, or smooth and wrinkled produce an intermediate form. Traits might be masked, but they never blended (*Figure 14*).

The reason for the specific frequency of the recessive trait in the F₂ generation can easily be accounted for by considering the simple permutations

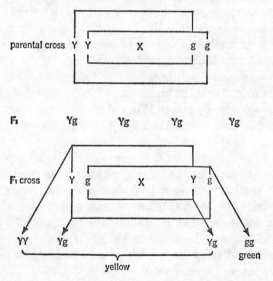

Figure 15. A cross of two pure strains of opposite type, again illustrating Mendel's law of segregation.

and combinations possible when two pure types are crossed. Each parent plant, quite obviously, will contribute a genetic unit, or gene, for the trait in question. Just as obviously, this must mean that the offspring will contain two units (one from each parent) for the trait. Different units of the same trait class—tall vs. short in the class of height, for example—are called *alleles*. Every plant contains two units for each trait, only one of which will be donated to a particular offspring. These units are donated at random. This means that either one has as good a chance as the other to turn up in the offspring. Now if we cross two pure strain plants of opposite type—a yellow with a green, for example— we can represent the cross as YY (yellow parent) × gg (green parent). All of the offspring will then be Yg, since all possible permutations and combinations yield four identical products (Yg, Yg, Yg, Yg). All the offspring will be yellow if the yellow allele is dominant over the green allele. Now if we cross any two of the F_1 plants (Yg × Yg) we should expect the following results: YY, Yg, Yg, gg. Twenty-five per cent of our offspring will be gg or green (*Figure 15*). In order to simplify the notation system in genetics, the letter symbolizing the dominant is used to represent both dominant and recessive alleles, the dominant in upper-case and the recessive in lower-case letters. Thus our F_2 cross would be written as follows: Yy × Yy. The result would be written YY, Yy, Yy, yy with the yy standing for the 25 per cent green-seeded offspring.

When the two units of a particular gene are identical (YY or yy), the organism is said to be *homo-*

zygous for that trait. When the two units are different alleles as in Yy, the organism is said to be *heterozygous* for that trait. Heterozygotes are also referred to as *hybrids*.

Independent Assortment

Mendel's experiments revealed another genetic regularity which he named the law of independent assortment. This law is concerned with the relationships among classes of traits (height and color, for example) rather than between traits of a single class. When Mendel crossed F_1 hybrids, he found that these various classes recombined at random in the F_2 generation. Thus, if you begin with only tall

	TY	Ty	tY	ty
TY	TYTY	TyTY	tYTY	tyTY
Ty	TYTy	TyTy	tYTy	tyTy
tY	TYtY	TytY	tYtY	tytY
ty	TYty	Tyty	tYty	tyty

= a total of 9 tall yellow plants:

4 TYty   2 TYTy
2 TYtY   1 TYTY

Figure 16. An example of Mendel's law of independent assortment, showing that class of traits (for example, height and color) become separated and rearranged during genetic recombination.

yellow and short green plants, you will eventually derive tall green and short yellow plants as well.

Tracing out such an experiment, we might begin with tall yellow, pure-strain plants and cross them with short green, pure-strain plants. This can be symbolized as TTYY × ttyy. In the F_1 generation, we should expect to end up with 100 per cent tall yellow hybrids. The permutations and combinations possible would yield only TtYy plants. Now, if we cross two hybrid tall yellow plants:

$$TtYy \times TtYy$$

we will end up not only with short green and tall yellow plants, but also with tall green and short yellow plants. Apparently the alleles for height and color become separated and rearranged in a process of genetic recombination. If they are unhooked and recombine at random, then we should expect the four types to emerge in the following frequencies:

9 tall yellow   3 tall green   3 short yellow   1 short green

If we begin with a large enough sample, this will indeed be the end result.

To understand how this recombination operates and why it results in the expected frequency, it is necessary to draw out the possible allelic contributions which each parent could theoretically make to the offspring. Obviously each parent can donate with equal frequency the following pairs of alleles (remember that each parent donates one allele for each trait): TY, Ty, tY, or ty. If these possible alleles are charted (*Figure 16*), it will be easy to determine the frequencies of the expected phenotypes. Note that all but the double-recessive (short green) type ttyy will have more than one genotype (or genetic configuration) in the phenotype. Thus the nine tall yellow plants will have the following genotypes:

TYTY (fully homozygous)
TYTy (homozygous for height, heterozygous for color)
TYtY (heterozygous for height, homozygous for color)
TYty (heterozygous for both traits)

Chromosomes

When one realizes that many thousands of traits contribute to the makeup of an organism, it should not be difficult to realize also the amount of potential variation present when genetic traits follow the law of independent assortment (*Figure 17*). There is, however, a restriction on the law which limits variation to a considerable degree. If all genes actually existed as independent units, like separate beads in a box, variation would be totally dependent on the laws of chance. Genes are, however, actually more like strings of beads, although, as we shall see later, neither analogy is particularly appropriate. At any rate, genes do not exist as totally independent units. They are linked together on chemical chains called *chromosomes*. Every cell contains several of these chromosomes, and their number depends on the particular species under consideration. Humans have forty-six chromosomes, and some species of fruit flies eight. The number of chromosomes is not dependent upon the size or relative primitiveness of the organism. Some one-celled animals (the protozoa), for example, are thought to have between forty and one hundred

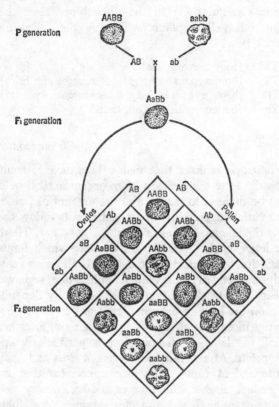

Figure 17. Another example of Mendel's law of independent assortment, in which strains of peas with yellow and smooth seeds and with green and wrinkled seeds are crossed. A and a represent yellow and green, and B and b smooth and wrinkled surfaces, respectively.

chromosomes. These genetic chains can be seen as a mechanism of continuity preserving, at least to some extent, stable combinations of genes, preventing them from completely sorting out during the reproductive process. As we shall see later, they do not prevent reassortment completely, but it should be obvious at this point that free assortment occurs only among genes on different chromosomes. Mendel was extremely lucky. Every one of the traits he used in his experiments occurred on a different chromosome, and therefore acted independently of every other unit. This is why they followed the law of independent assortment.

Codominance

But Mendel's luck was even greater than this. For not only can the law of independent assortment be violated by chromosome linkage, but the fact that genes are independent units (unit characters) can also *appear* to be violated, although this is actually not the case, at least in the chemical sense. Everyone knows, I think, that there are certain crosses in which it is possible to combine what appear to be two variant forms of a trait class. In certain species of plants, for example, it is possible to cross white- and red-flowered plants and produce an F_1 generation with pink flowers. When this occurs, the genes in question are said to be *codominant*. It is easy to prove, however, that even in the case of codominant genes blending of the genes themselves has not occurred. If the F_1 hybrids are crossed, the F_2 generation will contain white and red flowers as well as some pink ones. Let us see how this happens. We shall let R stand for Red and W for White (since there is no recessive gene, we shall not use a lower-case letter). The F_1 pink flowers will all be hetero-zygous RW. Now if we cross two of these we shall get the following results:

$$RW \times RW$$
$$RR\ 25\%\ (RW\ 25\%\ WR\ 25\%) = RW\ 50\%\ \text{and}\ WW\ 25\%$$

When we read these phenotypes it should be ob-vious that we end up with 25 per cent white, 50 per cent pink, and 25 per cent red flowers. There has been no blending of genes.

Polygenes

There is another case of blending inheritance, how-ever, which is more complicated. This kind of blending occurs when instead of one gene (and two or more alleles) for a trait, there are several genes at different places (*loci*) on a chromosome or even on different chromosomes. Such traits are said to be *polygenetic*.

Let us now imagine a case in which height is de-pendent upon three gene loci. Assuming that there are dominant and recessive alleles for each loci, a number of possible genetic combinations will pro-duce different heights in an F_1 generation which is the result of a hybrid cross. If the tall genes are dominant over the short genes, then the tallest in-dividuals would have at least one dominant gene at each locus. On the other hand, the shortest indi-viduals would be homozygous recessive at all three loci. Between these extremes there would be a dis-tribution or range of heights. The situation would be more complicated with codominant genes at each locus. In such a case the tallest individual would have to be homozygous tall at each locus (six tall alleles) and each short gene present would have some effect on total height. The existence of polygenetic traits is demonstrated through breeding experiments which show a range of variation within a particular trait class, providing, of course, that environmental variation is accounted for. In many instances, particularly in humans, with whom controlled breeding is impossible to achieve, it is difficult if not impossible to determine if a particular range of variation is due to a polygenetic effect or to environmental variation affecting the total phenotype.

While the polygenetic effect appears to violate the principle of unit characters, because a series of genes modify each other, it is not difficult to prove that the principle of unit characters is a universal phenomenon in genetics. This has been demonstrated again and again in breeding experiments in which masked genes reappear (i.e. show their effects) in the expected frequencies in the proper generation (*Figures 18a, 18b*).

Gene Modification

A further complication of the polygenetic effect results from the fact that some genes may not only have an additive effect, but can also modify or suppress the action of others. This phenomenon is known as *epistasis*. A gene which can suppress the effect of another gene at a different locus is said to be epistatic to that gene. (The suppression of a

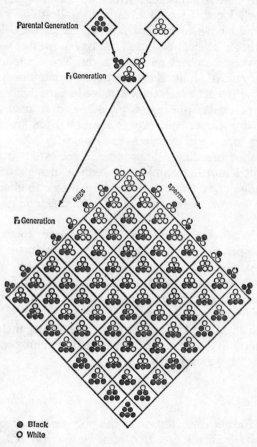

Figure 18a. Cross of black with white skin color, illustrating quantitative inheritance dependent on a six-allele codominant model.

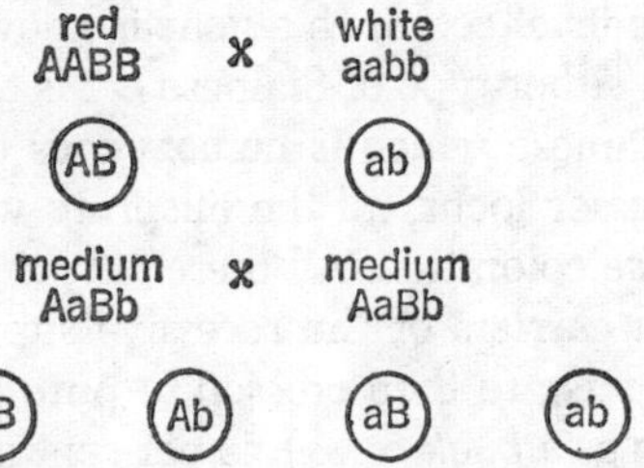

	AB	Ab	aB	ab
AB	AABB red	AABb dark	AaBB dark	AaBb medium
Ab	AABb dark	AAbb medium	AaBb medium	Aabb light
aB	AaBB dark	AaBb medium	aaBB medium	aaBb light
ab	AaBb medium	Aabb light	aaBb light	aabb white

Figure 18b. Cross between a wheat variety with red kernels with another having white kernels. This is another example of inheritance dependent on multiple genes.

recessive gene by a dominant allele at the same locus is a special case and is not called epistasis.) The suppressed gene is said to be hypostatic to the suppressor gene. A clear example of this effect is found in Siamese cats. The seal-point Siamese has a normal gene for dark color at one locus and a pair of recessive alleles at another which modify the full expression of that color. The blue-point Siamese has the same suppressor gene, plus still another gene at a different locus which modifies the color gene and

dilutes its effect. Such a gene is known as a *dilutant*.
When either type of Siamese cat is crossed to a pure
non-Siamese which is homozygous dominant at the
suppressor locus, all the offspring will display non-
Siamese coloration. All these offspring are, however,
hybrid carriers of the recessive suppressor gene. If
these cats are then crossed to pure Siamese (this is
known as a *back cross*) 50 per cent of their offspring
(in the F_2 generation) will be Siamese in coloration.
Furthermore, since the suppression depends upon a
recessive gene and therefore homozygosity, all cats
which look Siamese will produce 100 per cent
Siamese-looking offspring when bred to each other.
A non-Siamese × Siamese cross at the suppressor
locus would look like this:

PARENTAL GENERATION	SS × ss
F_1	Ss, Ss, Ss, Ss (no Siamese)
F_1 CROSS	Ss × Ss
F_2	SS, Ss, Ss, ss (25% Siamese)

Albinism in animals (white mice and humans, for
example) is usually due to the epistatic effect of a
suppressor gene on a normal color locus. The genes
responsible for color are hypostatic to the gene for
albinism.

It is also possible for genes at several loci to modify each other in such a way as to produce new variations of a trait. These combinations can be produced by genes which complement each other or
by what is known as the additive effect. A typical
case of the latter is provided by summer squash.
One type of sphere-shaped squash (genotype aaBB)
mated to another sphere-shaped squash (genotype
AAbb) will yield an F_1 generation of disc-shaped

hybrids (AaBb). The hybrid cross will produce nine discs, six spheres, and one fully recessive (aabb) elongate type.

It would be most convenient if the majority of genetic traits were caused by genes at a single locus, but this is not the case. The simple rules which Mendel discovered, however, remain as the basis for the study of variation. Each time a breeding experiment yields a significant departure from expected frequencies, this departure must be accounted for by a modification of theory. This modification must then be tested through further experimentation. This is how such phenomena as epistasis and the additive effect were discovered.

Sexual Recombination

Mendel's discoveries were based on the fact that in sexually reproducing species, hereditary material from each parent is combined in the genetic structure of the offspring.

Furthermore, it has been demonstrated that each parent donates exactly one half his or her normal chromosome number to each member of the succeeding generation. Since there are always two loci per trait, and since these loci are located along the length of the chromosome, it should be obvious that the full genetic complement of an individual is made up of homologous pairs of chromosomes. The alleles on these chromosomes may differ (they may be heterozygous) or they may be chemically identical (homozygous), but each set contains paired loci which control the same group of traits.

Sexual union produces a combination of genetic

material from each parent. Both the sperm of the male and the egg of the female carry one of the two homologues of each chromosome. These specialized sexual cells, or *gametes*, unite to form a *zygote*, which will, by virtue of the combination, contain the normally paired sets of chromosomes. In the human there are forty-six pairs of chromosomes, and consequently twenty-three single chromosomes in each gamete. Forty-six is the *diploid*, or zygotic, number; twenty-three is the *haploid*, or gametic, number.

The process of gametic formation and recombination in the zygote is one of the most important events in the evolutionary process, for it produces genetic continuity between generations. But each parent also provides a source of variation. We have seen, for example, that two parents with different combinations of traits can produce even further differences in their offspring through the recombination of trait elements. Tall green plants crossed with short yellow plants produce not only the parental types, but also short green and tall yellow plants.

If two populations of the same species are separated for several generations, long enough for genetic divergence to occur, and are then brought together, new combinations can be produced. It is possible that some of these combinations might have a higher selective advantage than pre-existing types. Even within a fairly stable population, sexual combination can produce a tremendous array of variation which can be played upon by selective forces. This type of variation and recombination requires no change in the basic chemical structure of the genes themselves, and is therefore quite different from chemical changes in the genes, a process which will be discussed later.

Gametogenesis

The formation of gametes (*gametogenesis*) is one of the most interesting processes of cellular biology. Basic to it is the reduction of chromosome number to one half the total complement in the parent. This process is known as *reduction division* or *meiosis* and differs, as we shall see, from the simpler process of ordinary asexual cellular reproduction or *mitosis*. Meiosis itself can be broken down into two major phases, each of which in turn consists of discrete stages. The first results in the duplication of chromosomes to double their number, and the second leads to the actual formation of gametes with a reduction to one half the total number of chromosomes.

The chromosomes are contained in the cell nuclei of all higher animals and plants, but only special cells located in the *gonads* or sex glands go through the process of meiosis and produce gametes.

Chemical evidence suggests that the process begins before visual changes actually occur in the nucleus. This stage is known as the *first interphase* and probably marks the time during which the chemical components of the chromosome are actually duplicated. The first observable stage of meiosis, the *first prophase*, is itself divided into substages, each of which marks significant events in the duplication and separation process. Prophase one begins with the *leptotene* substage. During the leptotene, the chromosomes appear to become long and thread-like. This is followed by the *zygotene,* in which homologous chromosomes pair so that alleles of the same locus come to lie side by side. In the next substage, the *pachytene,* the chromosomes become shorter and more condensed, probably due to coiling, although it is difficult to analyze microscopically exactly what happens to them. Toward the end of the pachytene, the chromosomes begin to duplicate themselves so that by the *diplotene* substage each original chromosome pair has become a *tetrad,* consisting of the two homologues and their copies. Sister *chromatids* (one homologue and its copy) are joined somewhere along their length by a small circular body known as the *centromere.* During the diplotene the tetrads begin to separate along the line dividing the original homologues. Each pair of homologue sister chromatids begins a migration to a side of the nucleus opposite their homologue pair. The completion of this migration, termed the *diakinesis,* ends the first prophase. In a series of stages known as *metaphase one, anaphase one,* and *telophase one,* the cell surrounding the active nucleus completes a division into two daughter cells. These daughter cells contain the original number of chromosomes, but the homologues will have become separated from each other. The new cells differ from the parent cell in that the total chromosome complement will be made up of the homologues from the original chromosomes, each with its exact copy or sister chromatid. In the end products of meiosis, the gametes containing one half the original chromosome number now begin to form through the separation of the sister chromatids into new daughter cells. This process begins in prophase two, passes through a second metaphase and anaphase, and comes to completion in a second telophase. During

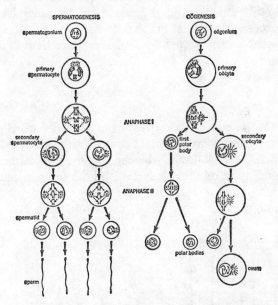

Figure 19. Traditional model of meiosis in the male and female animal. In the left diagram, the process of spermatogenesis results in the formation of four sperm; at the right, oögenesis results in the formation of one ovum and three polar bodies.

both the separation of chromatid pairs in the first division of meiosis and the final separation of sister chromatids in the second, the dividing units in the nucleus line up on a special structure known as the *spindle*. Under the microscope it appears as if fibers in the spindle become attached to the centromeres of each chromatid. Parental chromosomes do not necessarily line up on the same side of the spindle, so that a gamete may contain contributions from both the maternal and paternal chromosomes. This process contributes to the reshuffling of genetic material which occurs in sexual reproduction (*Figure 19*).

Crossing Over

The process of meiosis as it has been described so far reveals two important sources of genetic variation. The first occurs when homologue chromosomes end up in different gametes, and the second occurs when only one of four cells in oögenesis develops into a mature egg. While normal meiosis preserves

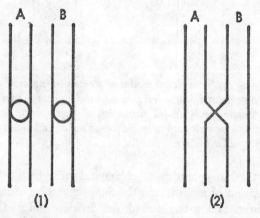

Figure 20. Crossing over: (1) homologous chromosomes each split into chromatids; (2) the crossing over and exchange between parts of homologues A and B.

all gene loci, the resulting gametes carry only one of two possible alleles for each trait. It is this separation of alleles in the gametes which leads to independent assortment. Variation is further enhanced during the first prophase by a phenomenon known as *crossing over* (*Figure 20*). This occurs when the four chromatids are lined up prior to separation. Chromatids are not rigid bodies and therefore tend to fall over one another on both sides of the centromere. When nonsister chromatids of the same tetrad cross over in this fashion, each strand may break away from its section attached to the centromere and join itself to the other chromatid at the crossover point or *chiasmata*. The process occurs reciprocally so that when part of one strand breaks and joins another chromatid, the second chromatid also breaks and donates an analogous part to the original chromatid. What crossing over amounts to is an exchange of genetic material between homologous chromosomes, and therefore a reshuffling of alleles. In a normal crossover the rejoined segments will continue to have specific loci at the original positions, but the linkage between traits will be changed. Thus, if in a set of homologues one contains an allele for tall at one locus and an allele for green at another, and the other homologue contains separate alleles for short and yellow, and if a crossover occurs between these loci, then the new arrangement will link tall and yellow on one chromosome and short and green on the other. The farther apart any loci are on a chromosome, the greater the chance that they will become separated during crossover. This fact has made it possible for geneticists to map the chromosomes with the approximate locations of specific genes. Such maps are known as *linkage maps.* Some chromosomes appear to be more resistant to crossing over than others. Those which are particularly resistant are referred to as *supergenes,* since they tend to act as a single genetic unit. This suggests that natural selection has operated to maintain advantageous configurations of alleles.

Crossing over does not change the over-all chemical composition of a chromosome, nor is any single gene changed. The induced variation is merely the result of a recombination of existing elements. It is interesting from the evolutionary point of view that the very existence of organized chromosomes rather than disorganized loose genes adds stability to the genetic configuration, but that counterforces operate to open this continuity to partial variation. Variation is enhanced by assortment in the formation of gametes and their combination in the zygote, and through the process of crossing over, which further limits the integrity of the chromosome as a single unit. It is perhaps even more interesting that counterforces act to restabilize certain genetic combinations through developed resistance to crossing over.

Mutation

Variation in genetic structure can also occur when genes themselves are altered chemically. This chemical process will be discussed in the next chapter, but I should point out here that such events occur. Any change in the gene or chromosome, whether mechanical or chemical, is referred to as a *mutation.*

Mechanical changes, such as crossing over, are referred to as *chromosomal mutations,* and chemical changes in single loci as *gene mutations.* While chromosomal mutations lead to new combinations, they do not actually change the basic chemical structure of the hereditary material. Their evolutionary significance results from the new combination of previously existing traits. Gene mutations, on the other hand, may add a new trait class to a species, or may significantly alter a trait class by producing a totally new form.

Geneticists have noticed that most gene mutations are disruptive to the successful adaptation of an organism. This is due, at least partially, to the fact that the species we study have been in existence for a considerable period of time and have therefore stabilized rather well to their environment. Any mutations away from the established form would be likely to have a detrimental effect. It is important to remember, however, that traits which are beneficial in one environment may not be beneficial in a new situation. Thus, when a species invades a new environmental niche, the genetic variation inherent in that species may produce new adaptive responses. The presence of some harmful recurrent mutations within a population may therefore be beneficial for the survival of a species under changing conditions.

Sex Linkage

Most sexually reproducing species are *dimorphic* (i.e. of two types), which are recognized as male and female. The determination of sex is a genetic

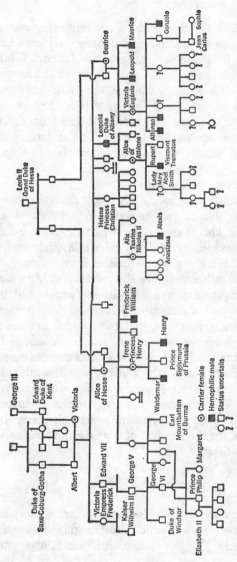

Figure 21. Genealogy of Queen Victoria and her descendants, illustrating the transmission of hemophilia, an X-linked recessive trait.

process based on chromosome differences. The chromosomes associated with sex are the only ones in the normal nucleus which are not completely homologous. One of the pair is shorter than the other, and for this reason they can be readily identified microscopically. In humans the short chromosome occurs only in males and is known as the Y chromosome. Female sexual chromosomes are both of normal length and are known as X chromosomes. When gametes form in the male, they are either X or Y. Since females can produce only X chromosomes, their gametes cannot determine sex. If this were *not* the case, a third type of offspring, YY, could be produced. In some species, particularly birds, the Xy complement produces females and the XX complement males. Since the Y chromosome is shorter than the X chromosome, it is possible for recessive traits located on the section of the X chromosome which has no homologue to express themselves in a single dose. Such traits are said to be sex-linked (*Figure 21*). Color blindness in humans is a good example of sex linkage. A man can be color blind with only one gene for the trait, but a female must be homozygous recessive to exhibit the trait. Since the gene for color blindness is relatively rare, the homozygous state is much less common than the heterozygous. But a male can only carry one allele at the color-blind locus. This is why most color-blind individuals are males. The following crosses should clarify this: Let C stand for the normal gene, c for the color-blind gene, and Y for the male chromosome lacking the C locus. If a heterozygous noncolor-blind female is mated to a normal male, we would expect the following results:

$$Cc \times YC$$

F_1 CY (normal male)   CC (homozygous normal female)
cY (color-blind male) and   cC (heterozygous carrier female)

In this cross only male offspring will be color blind.

If, however, a color-blind male (Yc) marries a female carrier (Cc), color-blind females will also result:

$$Cc \times Yc$$

F_1 CY (normal male)   Cc (carrier female)   cY (color-blind male) and   cc (color-blind female)

Sex-linked genes can be dominant as well as recessive. When this is the case, the results of the cross also differ from expected sex ratios in a normal Mendelian cross. If, for example, an Ay male is mated to an aa female, only daughters will carry the trait (Aa). All male offspring of such a mating will produce ay males. If the cross is between an Ay male and an Aa female, all of the daughters but only one half of the sons will exhibit the trait (*Figure 23*).

Sexual aberrations are never produced by YY combinations, but they can occur when the number of sex chromosomes is larger than the normal pair. This may result when some accident takes place in the meiotic process. In human beings, individuals triploid (XXY or XXX), tetraploid (XXXX, XXXY), and even pentaploid (XXXXY, or XXXXX) have been discovered (*Figures 22* and *24* show some examples). In every case normal sexual development is disturbed, and clinical symptoms appear in the affected individuals.

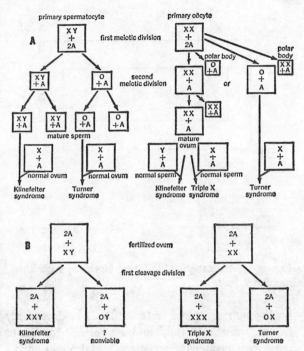

Figure 22. *Mechanisms by which aneuploidy of the sex chromosomes might develop. A. Nondisjunction in gametogenesis. B. Nondisjunction or chromosome loss in the zygote.*

Asexual Reproduction

Since I have been discussing genetic recombination in sexually reproducing species, it might be appropriate at this time to compare this process to *asexual* reproduction. Many lower organisms do not

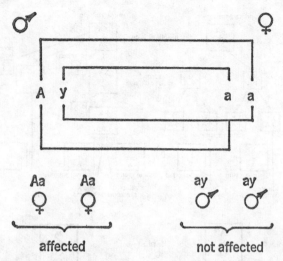

Figure 23. An example of the transmission of an X-linked dominant trait.

go through a diploid state, and others do so only rarely, sometimes only under stressful conditions when reassortment of genetic material might have a high survival value for the species. Reproduction of haploid organisms and the asexual reproduction of diploid animal and plant tissue are simpler than gametogenesis. Meiosis with reduction division is not necessary, and the process known as *mitosis* (*Figure 25*) merely consists of the reproduction of a single set of chromosomes prior to cell division. The process begins, as in meiosis, with the interphase in which the chemical constituents of the chromosomes are probably reproduced. In the prophase the chromosomes become visible and the orienting

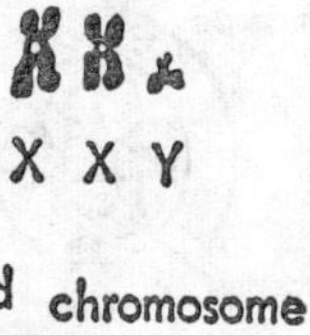

X X Y

23rd chromosome

Figure 24. The karyotype of a patient with the Klinefelter syndrome: forty-seven chromosomes and an XXY sex-chromosome constitution (shown here).

spindle begins to form. In the metaphase the chromosomes divide into chromatids, and these line up opposite their sisters on the spindle. In the anaphase the sister chromatids begin their migration to opposite poles of the spindle, which is completed in the telophase. The telophase also marks the reformation of the nuclear membrane which disappears early in mitosis, and the division of the cell into two daughter cells.

Since there is no crossing over during mitosis, and because gametes are not formed which then combine to produce new genetic combinations, asexual reproduction leads to a much higher degree of conformity in the offspring than does meiosis. Any genetic variation which occurs must be the result of either a chemical gene mutation or some aberration of the chromosome structure. Other things being equal, one would therefore expect genetic change to be slower in asexually reproducing organisms. This apparent stability is overcome, however, by the rapid reproduction rates of most of these organisms. The speed of division depends upon the particular species and external conditions, including availability of nutrients and temperature, but many species

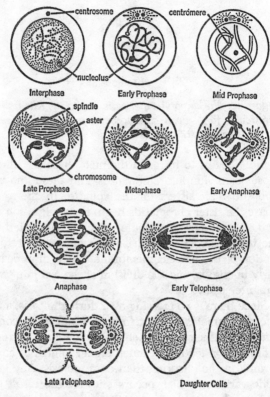

Figure 25. Mitosis in an animal cell having four chromo-somes.

reproduce several times a day under optimum conditions. In addition to this, since each division leads to two daughter cells, the rate of reproduction follows a geometric curve. A low mutation rate in such a rapidly reproducing species is generally sufficient to provide all the variation necessary for evolutionary change to occur. Further mechanisms of variation would probably raise the genetic load (the percentage of harmful genes) to a degree detrimental to the species.

Some organisms which usually reproduce asexually appear to have an alternate sexual mechanism which provides added variability under stressful conditions. These organisms have the best of both possible worlds. One of these sexual mechanisms, known as *conjugation,* is found in certain species of paramecium. If two strains, or mating types, of a particular species of paramecium exist in the same environmental niche, and if conditions for survival drop below the optimum by a certain degree, then conjugation is likely to occur. When this happens, organisms of opposite mating types pair and exchange nuclear material. When they separate later, each will have a new combination of genetic material in its nucleus. It is possible that some of these new combinants will have an increased viability.

Still another mechanism which releases hidden variation exists in these organisms. This process, known as *autogamy (Figure 26),* can occur within individual animals, and therefore does not require contiguity between organisms of opposite mating types. In autogamy the nucleus goes through a process similar to the first phase of meiosis. Chromosome pairs line up, duplicate, and then migrate to opposite sides of the nucleus, which then divides. The daughter nuclei migrate to opposite sides of the mother cell, which then divides. The new daughter cells have the original diploid number, and when they reproduce mitotically, this diploid number is maintained. (Gametes, therefore,

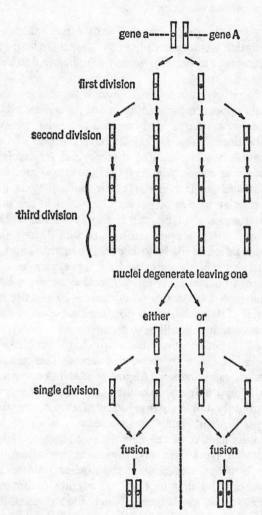

Figure 26. Distribution of a pair of genes in autogamy of paramecium.

are not formed.) Autogamy liberates variation, however, by creating homozygous diploids, for in the autogamous division sister chromatids migrate together to form the chromosome complement of the daughter cell. Any recessive genes masked by dominant alleles will be free to express their traits in those daughter cells, which after autogamy are homozygous recessive.

Chromosomal Variation

Another whole class of chromosomal mutations exists, which is quite different from crossing over, but which in some cases interacts with the crossover phenomenon. These mutations are all related to some form of noncrossover breakage of the chromosome, and may occur in both haploid and diploid forms, and in sexually and asexually reproducing species.

The first of these is a case of simple *deletion*, in which a piece of a chromosome is lost. Deletion is the result of breakage of a chromosome strand. Since a centromere itself cannot break, any separation of a chromosome strand will lead to two pieces, one with and one without a centromere. When cell division, either meiosis or mitosis, occurs, the piece without the centromere will be unable to line up on the spindle, and will therefore be lost in the daughter cell. The deletion will have three possible effects. (1) If, as is likely, the lost segment is crucial to organismic functioning, this type of mutation will be lethal, particularly in haploid organisms, where no homologue exists to maintain function. (2) If the deletion is not lethal to a haploid, it means that the daughter cell and all succeeding generations will display characteristics at variance with normal organisms. (3) In diploid organisms it means that recessive genes on the homologue which would normally be masked by the missing segment will appear in the phenotype. Breeding experiments with such animals will yield results at variance with the expected Mendelian frequencies only for those traits directly affected by the deletion.

Another type of chromosome mutation occurs when a segment breaks away from a chromosome and then rejoins it in the *inverted* position. Such *inversions* can lead to aberrant developments during meiosis, since homologue chromosomes will be unable to line up properly (*synapsis*) before the formation of tetrads. In haploid organisms there may be no effect on the phenotype unless the new position of the inverted genes affects their operation in the chemical system. This position effect will be discussed more fully in the next chapter.

In diploid organisms it is possible for a chromosome segment to become duplicated when, after a break occurs, a unit attaches itself to its homologue. In these cases some gametes will exist with deletions and others will have repetitive elements on the chromosome strand. Broken segments may also attach themselves to the wrong chromosomes (*translocation*). Finally, the meiotic process may be partially disrupted so that a gamete ends up with the diploid rather than the haploid number for certain chromosomes (*Figure 27*). When these gametes unite with normal gametes to form a zygote, the chromosome number for these particu-

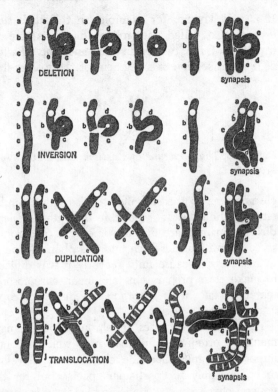

Figure 27. Examples of types of chromosome rearrangements.

lar chromosomes will be triploid, and if two diploid gametes unite, the number in the zygote will be tetraploid. All of these events (duplication, inversion, deletion, translocation, and polyploidy) are generally harmful to the resulting offspring, except in plants in which the occurrence of polyploidy is quite common and results often in increased vigor.

It is now known, for example, that *Down's syndrome*, also called Mongolian idiocy, is due to triploidy of the twenty-first chromosome (*Figure 28*). Translocation of a segment of the twenty-first

21

Figure 28. The karyotype of a patient with Down's syndrome.

chromosome to the seventeenth also has a similar effect in humans. It is possible that other genetic diseases in man are due to some kind of chromosome aberration. Research into this phenomenon in man is new, because until recently it was difficult to make clear preparations of human chromosome material (known as a *karyotype*) for microscopic examination.

Again, while changes in chromosome arrangement are often harmful, they may well have had important significance for evolutionary development, not only for particular species but for the general chemistry of genetics, which itself is an end product of a long period of evolutionary development.

All of the genetic mechanisms discussed in this chapter have some effect on the continuity or variation of hereditary material. Some of these effects have already been mentioned, but I should now like to discuss their evolutionary significance.

It should be obvious that in heterozygote loci, dominant genes buffer recessive genes by suppressing their phenotypic effect. Codominance results in a modification of the expression of either gene. If a particular recessive gene is lethal or less adaptive than the dominant, it would if unmasked tend to drop rapidly out of a population through the process of natural selection. The number of recessive genes maintained within a population is actually very large, but most of them are protected from selection pressures by the dominant gene. Epistasis also protects variation by suppressing gene action.

If all genetic variation were expressed in the phenotype, the vitality of the species would be lowered in most cases. The reason is that selection operates to improve goodness of fit between the phenotype and the environment. It would appear that time has produced an evolutionary compromise in which phenotypic variation is much lower than genotypic variation. Critical periods in the life cycle of a population may then be responded to with some of the stored genotypic variation.

III
CHEMICAL GENETICS

Every living organism is a self-regulating system. This means that a definite structure exists which consists of interlocking and interacting units capable of carrying out necessary processes and of maintaining these processes within a specific normal range. In addition to maintaining the activities of the system, the organism must build and repair these units. Evolution has produced dynamic systems of great complexity. Even the "simplest" organisms must complete an astronomical number of chemical reactions merely to maintain their structural integrity. The metabolic process, for example, which converts materials absorbed from the environment into energy and necessary cellular structures requires several thousand enzymes for its successful operation. The absorption and synthesis of basic material includes the uptake of simple chemicals which are combined into more complex molecules, as well as the uptake of complex molecules which must be broken down to be rebuilt according to a specific architecture. The absorption of these chemicals must be a highly selective process, and their resynthesis must proceed according to a determined sequence.

Each individual living system has a finite existence. The continuation of a species depends upon reproduction. While single-celled organisms need only reproduce exact copies of themselves, multi-cellular organisms face additional problems. The uniting of two gametes is only the first step in the formation of any complex animal or plant. From worm to man, what begins as a simple cell capable of replication through mitosis must differentiate into the specialized cells which constitute and control the organs of nervous and muscular function, circulation, absorption, and excretion. The architect and engineer of organismic structure is the hereditary material which controls the heritage of a species and directs the operation of each individual organism.

DNA

The major hereditary unit is a long *polymere* (or chemical chain) known as *deoxyribonucleic acid* (*Figure 29*). Its core consists of sugar-phosphate molecules linked together in parallel chains. A series of bases is attached to one side of each chain. These are *adenine, cytosine, thymine,* and *guanine* (*Figure 30*). While the sugar-phosphate links are identical in structure, the sequence of bases may vary along the DNA molecule. These bases serve as connecting units between the two sugar-phosphate chains. The base sequence on one chain controls the base sequence on the other, because the hydrogen bonds which unite the base pairs can occur only between thymine and adenine, on the one hand, and cytosine and guanine on the other. If one chain contains the sequence adenine, adenine, adenine, thymine, the other chain must contain the opposite sequence—thymine, thymine, thymine, adenine. If

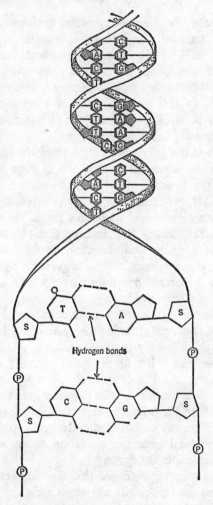

Figure 29. A schematic diagram of the Watson-Crick model of the DNA molecule.

one chain has adenine, cytosine, guanine, the opposite chain will have thymine, guanine, cytosine, and so on. The two sugar-phosphate chains united by hydrogen bonds at the base pairs make up the complete DNA molecule. Structurally this molecule appears to coil around itself in a double helix. Models of DNA have a superficial resemblance to the common pond algae *Spirogyra*.

DNA is the primary constituent of chromosomes and is now accepted as the basis of gene activity. The major function of genes in the nonreproducing cell is the production of enzymes which regulate cellular structure and function through their initiation and control of chemical reactions. Gene activity also controls the differentiation of structure which occurs in multicellular organisms. Again this process is carried out through enzymatic activity. Genes not only produce these enzymes, they also regulate their production so that only appropriate reactions occur at specific times. This control mechanism, which is only partially understood, is important because it lies at the basis not only of self-regulation, but also of differentiation in organisms all of which contain exactly the same genes in every cell. It is this regulatory mechanism which makes it possible for an undifferentiated zygote to develop into the total organism made up of an array of highly specialized organs.

There is strong evidence that the arrangement of bases on the DNA molecule act as code units in enzyme synthesis. Adenine, thymine, cytosine, and guanine are the four letters of the code. These letters are used to spell out any one of twenty amino

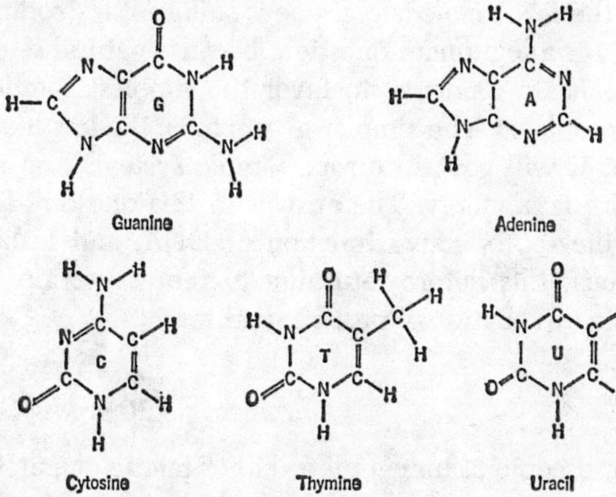

Figure 30. The bases of DNA and RNA (uracil being substituted for thymine in RNA).

acids used in the synthesis of proteins. If we assume that the code uses the smallest possible number of letters to spell out the entire twenty-word vocabulary, we shall find that only three letters are necessary for any one amino acid. A consideration of all the possible permutations and combinations of the four letters into three-letter words or triplets yields a total of sixty-four. Some amino acids can be coded by more than one triplet and some triplets spell out nonsense words, which may act as "punctuation marks" separating genetic messages from one another.

The active coding of amino acids comes off only one of the two sugar-phosphate chains. The other chain does not participate in the production of enzymes. This should immediately raise the question of why there should be two sugar-phosphate chains on the DNA molecule, if one is sufficient for coding. This is a legitimate question, because natural selection has a tendency to favor the simplest solution to problems. The simpler a system is, the less likely that it will contain errors. Simple systems also require less energy. The answer to this question lies in the reproductive function of DNA, and I shall discuss this before returning to the as yet incomplete explanation of protein synthesis.

Chromosome Reproduction

Microscopic examination of either miotic or mitotic cells reveals a duplication of chromosomes in the

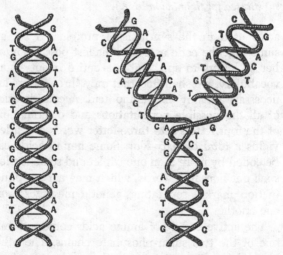

Figure 31. The double-helix model of the gene and chromosome structure (left) *and of its replication* (right) *as suggested by the Watson-Crick hypothesis.*

nucleus. Since DNA is the major element in these structures, it would appear that DNA itself is duplicated during the process of cell division. If one measures the quantity of DNA prior to the reproductive phase and again before cell division is complete, it is found to be double the original quantity. Thus cytological and chemical evidence both point to the duplication of DNA as a major step in reproduction. Looking again at the DNA molecule, let us imagine that prior to reproduction it uncoils and separates into two chains. Each chain will now consist of sugar-phosphate links plus one set of bases arranged lineally down the molecule. If the nucleus is able to produce new single sugar-phosphate links with attached bases, the separated but still complete single strands of DNA could function as templates for the formation of sister chains. Each old DNA strand could combine with a new strand to reconstitute a complete molecule. The two original strands of the old DNA molecule would then give rise to two complete DNA molecules identical in structure to the original (*Figure 31*). Sophisticated experiments have demonstrated that this is very likely what happens in both mitosis and meiosis. Sister chromatids appear to be the result of this process. The double strands of DNA are powerful mechanisms of continuity in the reproduction of like units.

Protein Synthesis

DNA is nuclear material, and although it participates in the synthesis of cellular substances, it must do so in combination with another related molecule which is able to transmit messages from the chromosomes in the nucleus to the cytoplasm of the cell. This substance, *ribonucleic acid* or *RNA*, occurs in at least two forms, each of which has a particular function in the protein-building process. Like DNA, this molecule contains four bases linked to a sugar-

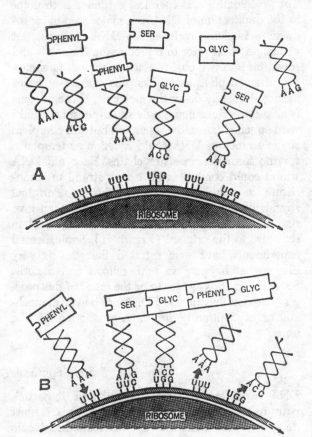

phosphate chain, but the base *uracil* is substituted for thymine (*Figure 30*). RNA is not implicated in the reproduction of genes, and it need not occur as a double chain. The fact that RNA occurs as a single strand is further evidence for the specific rela-

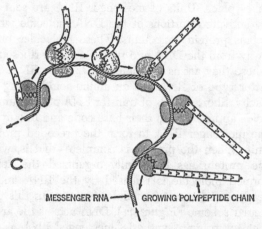

MESSENGER RNA ⟶ GROWING POLYPEPTIDE CHAIN

Figure 32. The messenger RNA (A) contains a sequence of base pairs complementary to those of the DNA strand that it "copied." The newly formed messenger RNA then moves from the nucleus of the cell to the cytoplasm where it attaches itself (B) to a ribosome. The transfer RNAs (C), joined to their particular amino acids, gather along the messenger RNA on the ribosome. The order in which they assemble is presumably dictated by a matching of the sequence of bases along the ribosome and in one area of the transfer RNA molecule. As the amino acids form a peptide chain, the transfer RNA molecules are released. The same messenger RNA (C) molecule is usually "read" by a number of ribosome units simultaneously. (From Curtis)

tionship between double-strand structure and the reproduction of genetic units.

One form of RNA, known as *messenger RNA*, carries a DNA code message through the nuclear membrane to special areas in the cytoplasm known as *ribosomes*. It is at the ribosomes that synthesis takes place. Units of messenger RNA are sent out by activated portions of the DNA molecule which act as protein code units. These molecules break away from the DNA and pass out to the ribosomes, where they act as templates for *transfer RNA*. The latter type of RNA has an amino-acid-bearing capacity. Short strands of transfer RNA pick up amino acids appropriate to their base code and line up on the messenger RNA to form the precoded protein unit. When the process is complete—that is, when the protein has been fully organized—the newly formed molecule is released by the RNA and is ready to take part in cellular activity. In this particular scheme (*Figure 32*), DNA acts as the architect for protein synthesis, messenger RNA as the blueprint, and transfer RNA as the construction engineer.

TABLE 1. The twenty essential amino acids with their triplet codes. UAA or G spell gaps and act as punctuation. Note the several redundancies. (Commas separate single substitutions.)

1. ALANINE — GCU, C, or A	5. CYSTEINE — UGU, or C
2. ARGININE — AGA, or G CGU, or C, A, G	6. GLUTAMIC ACID — GAA, or G
3. ASPARAGINE — AAU, or C	7. GLUTAMINE — CAA, or G
4. ASPARTIC ACID — GAU, or C	8. GLYCINE — GGU, or C A, G
	9. HISTIDINE — CAU, or C

10. ISOLEUCINE — AUU, or C
    AUA

11. LEUCINE — CUU, or C,
    A, G
    UUA, or G

12. LYSINE — AAA, or G

13. METHIONINE — AUG

14. PHENYLALANINE —
    UUU, or C

15. PROLINE — CCU, or C,
    A, G

16. SERINE — AGU, or C

17. THREONINE — ACU, or
    C, A, G

18. TRYPTOPHAN — UGA, or
    G

19. TYROSINE — UAU, or C

20. VALINE — GUU, or C, A,
    G

Gene Mutations

Gene mutations, which actually change the structure of the synthesized protein molecule, are thought to be changes in the basic code unit of DNA. If only one code word of a DNA message is changed, the resulting protein will be significantly altered. If the alteration occurs on an active site of the protein— that is, a portion of the protein which enters into chemical reactions—the change can effect an entire enzymatic process and therefore alter any phenotypic traits which are the outcome of such activity.

A particularly well-documented example of a mutation on the chemical level is provided by the fatal disease known as *sickle-cell anemia*. Normal hemoglobin contains 574 amino acids arranged in two sets of paired units known as *beta* and *alpha* chains. In normal hemoglobin, the sixth amino acid in the beta chain is *glutamic acid*. *Figure 33* shows how a single change in the DNA code can cause a substitution of *valine* for the glutamic acid. This leads to abnormal *hemoglobin S* and the pathological condition sickle-cell anemia (*Figure 34*). The change in hemoglobin structure affects the oxygen-

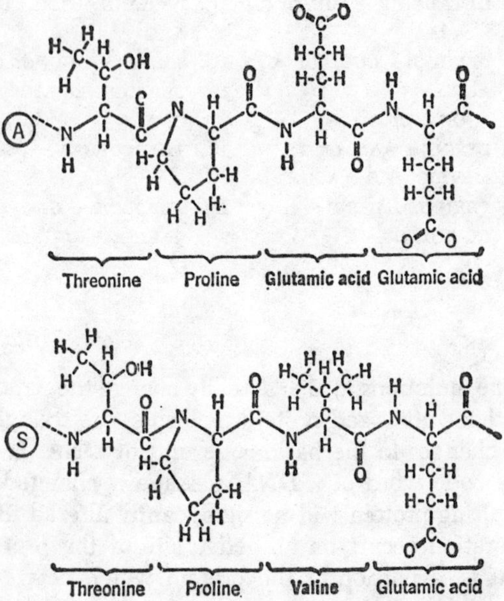

Figure 33. Corresponding sections of the molecules of hemoglobins A and S. The glutamic acid in A is replaced by valine in S.

transporting function of the red cell. The sickle-cell gene acts as a partial recessive, and although heterozygotes show some effects of the disease they are able to live, reproduce, and act as carriers of the mutant gene.

The genotype for homozygous sicklers is written as follows: $\alpha_2{}^A\beta_2{}^S \cdot \alpha_2{}^A\beta_2{}^S$. The $\alpha_2{}^A$ stands for two normal alpha chains. The letter A denotes normal adult hemoglobin. The $\beta_2{}^S$ stands for *two* abnormal beta chains producing hemoglobin S. The form $\beta_1{}^A\beta_1{}^S$ cannot occur. The genotype for the hetero-

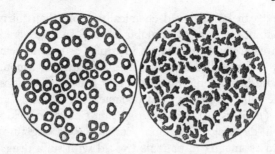

Figure 34. Phenotype produced by sickle-cell hemo-globin (Hb S), on the right, as compared with normal erythrocytes, left.

zygote is written as follows: $\alpha_2{}^A\beta_2{}^A \cdot \alpha_2{}^A\beta_2{}^S$. The individual in this case receives the genetic complement for normal hemoglobin from one parent and the abnormal complement from the other. Finally, the genotype of a homozygous normal individual is written as follows: $\alpha_2{}^A\beta_2{}^A \cdot \alpha_2{}^A\beta_2{}^A$.

The structure and genetics of the hemoglobin molecule are particularly well known, and several mutations have been noted for humans. Some of these involve active sites (the sickle-cell mutant is one example) while others appear to have no effect on normal function. Among the changes are the following: On the alpha chain, the fifth amino acid is normally alanine. A change in triplet code from GCC (or U) to GAC (or U) will change the amino acid to aspartic acid. On the beta chain, a change in the sixty-seventh amino acid (which is normally valine) to glutamic acid occurs with the triplet change GUA (or G) to GAA (or G). Note that in both cases a single base substitution leads to a change in amino acid and therefore to a mutation that can be analyzed chemically. More than thirty mutant forms of hemoglobin have been discovered. In each case, the change in triplet code is known, as is the particular amino acid substitution. Among those hemoglobin mutations that cause pathologies, the following have been documented: Hemoglobin C (β–6 lysine) causes an anemia similar to sickle-cell disease; hemoglobin Zurich (β–63 arginine) causes anemia in patients treated with sulfa drugs; hemoglobin M which occurs in three forms (α–58 tyrosine, α–63 tyrosine, and α–67 glutamine) interferes with oxygen transfer.

Comparisons between the structure of the hemoglobin molecule in various species allows geneticists to study evolutionary distance. This provides an additional method and a check on the more traditional methods of classification according to morphological differences and similarities. This type of analysis can be carried out on other chemical constituents of living systems as well.

Proteins are complex molecules built up from sequences of amino acids. Their activity depends upon the order of specific amino acids and the actual structure or architecture of the molecule which consists of one or more chains folded into characteristic patterns. These molecules have specific *active sites* that enter into chemical reactions. Mutations (amino acid substitutions) may or may not occur at active sites. Those that do will have an immediate effect on the metabolic activity of the protein. Those that do not affect active sites may accumulate as *silent mutations*. A series of silent mutations may build up in a molecule until a further mutation, acting with the previous silent changes, produces a significant alteration of the molecular structure and hence its activity. The analysis of hemoglobin has revealed several silent mutations.

Changes in the chromosome through deletion and duplication also alter the chemical composition of the total DNA strand, and so it is not surprising that such aberrations have their effect on the organism. Less obvious is the fact that changes in the position of a segment of a chromosome through inversion or translocation can also affect gene activity. This fact strongly suggests that the behavior of any one segment of DNA can be influenced by associated parts of the chromosome. Thus, even on the molecular level one could say that environment is intimately related to the operation of heredity.

Recent experiments carried out with *thermophilic* organisms (those which can grow at abnormally high temperatures) inhabiting hot springs have demonstrated that external environmental effects can alter the action of DNA.

The DNA molecules of these algae are able to produce *thermostable* proteins—that is, proteins which are heat-resistant and therefore do not denature at relatively high temperatures. The ability of the genetic structure to adjust to such an environment is itself an amazing demonstration of natural selection and the range of possible mutations. From the environmental point of view it is interesting that if the DNA from these organisms is cooled to a temperature approaching the more normal range for living species, the same code produces a different set of proteins. This is clearly not a case of mutation, for there is no change in the structure of the DNA. If these molecules are brought back to their original temperature they will produce the original set of proteins. The environment, in this case temperature conditions alone, has a direct effect on the action of DNA. Since all chemical reactions require energy for their operation, it is likely that significant changes in temperature affect the chemical procedures by altering energy levels. It is important to note that the effect of temperature on the operation of protein synthesis in thermophilic algae is not the same as that encountered in the coat color of Siamese cats. In the case of these cats, it is the coat-color pigment itself which is temperature sensitive. Temperature changes in this case do not have a direct effect on the operation of DNA.

Thermophilic bacteria and algae are rather specialized organisms, but it is possible that environmental stimuli might alter the effect of DNA in more common organisms.

It now appears that the *internal* environment of the organism on both the extra- and intracellular levels has very important effects upon DNA operation. In fact, these effects provide the clues for the problem of selective activity in all organisms and cellular differentiation in complex animals and plants.

Enzyme Production and Feedback

Basic to an understanding of DNA regulation is the concept of feedback or self-regulation mentioned at the beginning of this chapter. In a feedback system, information generated as a by-product of ongoing process has a regulatory effect on that process. One of the simplest forms of a feedback device is the governor of a motor. If a set of expanding metal bands with counterweights at their centers are attached to a shaft, the speed of which is dependent upon the motor, acceleration will tend to drive the bands farther and farther apart by centrifugal force. If the weights are adjusted so that they exert a drag sufficient enough to slow the motor to a predetermined rate, minor fluctuations of power will not alter the rate of motor revolutions. Centrifugal force of the governor will exert accelerating pressure when the power source weakens temporarily, and decelerating pressure when the power increases. Of course, a major change in power will override the ability of the feedback mechanism to operate successfully. When the power is turned off, the motor will eventually come to rest. Feedback systems can maintain the integrity of a system only within certain limits. Some of these systems are more efficient than others in their system maintaining abilities.

Another common feedback mechanism is the thermostat in modern heating units. A temperature-sensitive bar is located in a switching mechanism which controls the furnace. As the temperature rises, the bar bends away from the point of electrical contact and breaks the circuit. This turns off the furnace. When the temperature drops sufficiently, the bar re-establishes contact and the furnace begins again. The distance necessary for the bar to travel before it breaks contact can be adjusted so that greater or less heat can be generated by the furnace. If the bar setting is calibrated to a specific temperature, the thermostat will regulate the production of heat within a narrow range of the setting. Again, the system which the feedback mechanism maintains can be destroyed. In this case if the furnace runs out of fuel, the thermostat will obviously be unable to maintain the temperature setting.

Biological feedback mechanisms of varying degrees of complexity are common, for they form the basis of organismic integration. The control of thyroid hormone secretion is a good example of this activity. The thyroid gland produces rather large quantities of *thyroxin* which are stored in the glandular tissue. Secretion of this substance is stimulated by *TSH*, a thyroid stimulating hormone which is produced in the *pituitary gland* or *hypophysis*. TSH is released into the system as a response to metabolic needs which are conveyed to the hypophysis from the *hypothalamus* (the portion of the brain which controls body temperature and respiration). The TSH is conveyed via the bloodstream to the thyroid, where it triggers the release of thyroid hormones—again into the bloodstream. These hormones influence the metabolic activity of the body, but at the same time they produce an inhibitory effect on the secretion of TSH. Thus they exert a dampening effect on the system until such time as renewed stimulation is required.

The feedback mechanisms which appear to be involved in DNA operation are of two types, which are called *extracellular* and *intracellular*. Intracellular control operates both in the regulation of normal cellular processes and in differentiation, although the latter is probably dependent on a combination of intra- and extracellular reactions. The first process which I shall discuss here is that concerned with reaction stimulation.

It would be uneconomical for the cell to produce its total vocabulary of proteins at a constant rate. Instead, certain reactions are stimulated only at their proper time. Many of these appear to be triggered by the presence of particular *substrates* (a substance to be broken down in enzyme activity). Experiments with bacteria have demonstrated that enzymes involved in the breakdown of lactose (a sugar used in cell metabolism) are produced by the cell only in the presence of the substrate (the lactose). Actually, two enzymes are necessary for this aspect of cell metabolism. One, a *permease*, makes the cell membrane permeable to the sugar, and the other, *galactosidase*, is actually involved in the breakdown reaction.

It is possible to produce mutations in these bacteria which affect one, two, or all three of the phenomena associated with this particular process. One mutation can eliminate the permease, another the galactosidase, and still another can stimulate the production of enzymes even when the substrate is not present. The latter possibility suggests that the feedback mechanism itself is under the control of a specific gene which can be altered through mutation. Two French scientists, François Jacob and Jacques Monod, who have worked on this problem, have suggested a model of gene activity which fits well with the process of *induction* (the stimulation of enzyme production in the presence of a substrate). The authors suggest that three separate genes are involved in induction. These they call the *structural gene* (which is responsible for enzyme production), the *regulator gene* (which produces a substance inhibitory to enzyme production), and the *operator gene* (which triggers the activity of the structural gene). The hypothetical sequence of events in the production of galactosidase occur in the following sequence: When there is no substrate present, the regulator gene secretes its inhibitor, which blocks the operator gene suppressing enzyme production. In the presence of substrate or some other substance sensitive to its presence, the regulator gene is inactivated. Chemically this can occur when the repressor molecule combines with the inducer. This frees the operator gene, which then stimulates the structural gene to perform its normal activity. A mutation at the site of the regulator gene can destroy its ability to inhibit enzyme production. When this occurs, enzymes are produced either with or without the presence of substrate. The regulator gene appears to be on a different chromosome than the operator and structural genes, which, at least in the experimental case, are closely linked. When these two units are separated—as, for example, in crossing over—the chemical productivity of the structural gene is destroyed. Thus, while the regulator gene need not be adjacent to the gene which it suppresses, the operator and structural gene make up a linked unit which is known as the *operon*.

Some feedback systems in enzyme production work like the model presented above for TSH—thyroid activity. That is, the production of the enzyme itself has some effect on the producing gene, so that at a certain level of concentration the production of enzymes is inhibited. The two inhibitory mechanisms described here serve different functions. The first is an economic device which does not allow the cell to produce a substance until it is needed; the other inhibits overproduction.

One of the major extracellular regulating mechanisms which is just beginning to be understood is hormone control on gene action. Hormones are substances which themselves act as enzymes in the organ system of an individual. Sex hormones are involved in the development of secondary sexual characteristics such as body hair and voice changes in the male. Thyroid hormones are closely linked to the control of metabolism and growth, as are some of the secretions of the hypophysis. Most but not all hormones are produced by the endocrine or ductless glands and are carried to all parts of the body by the bloodstream. Most hormones have specific target organs which they affect even when secreted in small doses. They can also stimulate or inhibit action in other regions of the body, particularly when they occur in abnormally high concentrations. Recent evidence suggests that hormones also act as messengers to specific cells, triggering or inhibiting gene action and therefore enzyme production.

The fact that specific gene action can be inhibited or stimulated under particular circumstances presents a clue to the problem of cellular differentiation. The genetic code of multicellular organisms must have some sort of time mechanism built into it which stimulates complex chains of events in a precoded sequence. In addition to this, as the process of differentiation unfolds, materials will be produced at the specific sites which have a further selective effect on the individual parts of the total structure. Each cell of the body contains the same chromosomes and genes which have been copied over and over again in the reproductive process, but in some way each cell also functions according to its special role in the total organ system. Each cell contains the master code system which determines cellular, and consequently, organ specificity.

Evolution has produced a series of complex systems which not only build themselves out of undifferentiated material drawn from the environment, but also manage to form specialized structures from a single basic unit. DNA, which provides the basic plan for all living forms except for some viruses which contain only RNA, has the same basic structure whether it is in the nucleus of bacteria or man. The major evolutionary event in the beginning of life was the development of a self-replicating substance which reproduced through the utilization and reorganization of chemical substances from the environment. If the DNA molecule were completely stable, evolutionary change could not occur. The important thing is that while its basic structure (the sugar-phosphate chain) is stable, the series of base pairs may be altered in an almost infinite variety of ways. It is the accidental reshuffling of these base pairs which provides the basis of change. As a code system, DNA spells out only twenty words, but these words can be combined into an infinite variety of sentences. It is these sentences which constitute the messages for the production of essential proteins. These proteins are themselves units in long chains of chemical events which in combination with environmental factors produce phenotypic traits. The variation produced by chemical mistakes in the coding process are sorted out in particular environments by the process known as natural selection. Most of these mistakes have unfortunate consequences for those organisms in which they occur, but from time to time they provide the basis for a better-adapted system.

Pleiotropy and Penetrance

It must be stressed that DNA produces proteins and not phenotypic traits, at least not in the direct sense. A single gene may affect not one but a series of traits, which can be called a *trait complex*. This is true because a single protein may lie at the base of various chemical pathways which themselves produce a host of phenotypic responses. The multiple effects of a gene are called *pleiotropy*. Most, if not all, genes are pleiotropic. The disease phenylketonuria referred to in the first chapter provides a good example of the pleiotropic effects of a gene. The direct effect of the mutation responsible for this disease is merely the absence in the system of a specific enzyme, *phenylalanine oxidase*. This leads to the excretion of phenylpyruvic acid in the urine. The pleiotropic effects of this enzyme deficiency include a severe form of mental impairment and a pigmentary disturbance which produces light hair. The genes responsible for differences in blood groups in man—the ABO (H) system—appear to be related to the statistical incidence of different diseases. Individuals with blood group A have a higher incidence of duodenal cancer than individuals with other blood groups. On the other hand, blood group O appears to be related to the incidence of peptic ulcer.

The phenomenon of pleiotropy has consequences for the study of natural selection. The selective value of a gene may depend on one or several of the phenotypic traits associated with it. Thus, while it is possible to determine the selective value of a particular trait in a population by comparing the fertility rates of individuals exhibiting the trait with those who do not, the derived value may actually depend upon some other activity of the gene responsible for the trait in question. Traits are markers for genes, and what is actually measured in selection studies is the selective value or coefficient of selection of a gene.

Another important point to consider as far as selection is concerned is the interaction between heredity and environment in the expression of a gene. Depending upon conditions which may be either external or internal, a particular gene may be expressed along a continuum ranging from a total lack of expression to maximum effect. The degree of expression of a genetic trait is referred to as *penetrance*. If for some reason a normally harmful gene has low penetrance, then its effect on the survival of a population will be less marked. Again it is necessary to stress that genes act in combination with environmental pressures and not as isolated determinants. The genetic aberration which produces diabetes, for example, might have a penetrance near zero in a population whose diet was extremely low in carbohydrates. The presence or absence of the diabetic gene in such a population would have little effect on fertility and could not be said to have a low selection coefficient. If the same population changed its eating habits, however, the combined effect of the gene plus the environment might reveal a rather high negative selective value for the gene. The introduction of insulin therapy into the same population at a later date would change the situation again, creating a more neutral status for the gene.

The combined effects of penetrance and pleiotropy help to make selective value a flexible proposition. Environmental factors might well suppress or modify one action of a gene and have little effect on some other. Any statement relative to the selective value of a gene must therefore always be given in terms of a specific environmental context.

IV

POPULATION GENETICS AND EVOLUTION

Our examination of genetics has been limited thus far to the simple rules of inheritance and to a consideration of the chemical basis of heredity. The evolutionary process cannot be understood, however, without analyzing the effects of genetics and environment as they interact in breeding populations.

Population Models and Probability

The theory of evolution implies that beneficial changes in genetic structure can be preserved in living forms and passed on through the reproductive process to the next generation. This means that within any species there must be a certain rigidity which operates to preserve evolutionary gain and a certain flexibility which provides the material for further adaptation. To speak of the rigidity or flexibility of a species, however, one must refer to a group within which there is a sharing of genes through the mechanism of sexual reproduction. Most species cannot be considered in this way, for they themselves are made up of separate spatially distributed breeding populations. The boundaries between these populations can be considered to be the boundaries between distinct gene pools. This does not mean that breeding is totally confined within such groups, for some gene flow must occur if the species is to be maintained as a genetic entity. It is convenient and necessary, however, to construct a simple model of a breeding population which does not conform totally to any actual situation. This allows us to project more realistic and complicated problems against a set of general abstract principles. Another necessary assumption of the model, which is rarely if ever achieved in natural populations, is that of random mating or *panmixia*. This means that in any class of possible events one event is just as likely to occur as any other in that class, or that in a sample of objects from which choices are to be made one object has as good a chance of being chosen as any other in that sample. In mating, as we have already mentioned, this means that any individual member of a population has as good a chance as all other members of the same sex to mate with any other opposite-sexed member of that population.

An abstract model of a population with the two characteristics described can be manipulated mathematically by introducing variables at first one at a time and then in different combinations to yield sets of expected results. All of this can be done with paper and pencil, using simple arithmetic and algebra. The results can then be tested either experimentally in the laboratory by introducing predetermined variables or through the observation of natural populations within which a series of known events is occurring. A high correlation between predicted and observed results tends to confirm the particular hypothesis concerning the model under test. A low correlation, on the other hand, demands an explanation which may well lead to a modification of theory. In this type of analysis it is necessary to have objective criteria for judging the significance of the result: in other words, what degree of correlation is necessary for confirmation, what constitutes high or low correlation? This factor is itself variable, since it depends upon the size of the experimental sample and the number of variables tested. According to simple Mendelian rules, for example, the result of a simple hybrid cross should yield a 25 per cent return to the recessive phenotype. In the actual experimental situation, however, the expected result will rarely occur. The experimenter must determine whether or not variation away from the expected outcome is due to chance factors alone or to some previously undiscovered factor.

Statistical manipulations exist which allow the experimenter to determine the significance of his result. These tests tell him whether observed deviations are due to chance factors which do not actually affect the variables or to undiscovered variables which must be found and explained. The usual procedure in these matters is to test the so-called *nul hypothesis,* which states that variation is due to chance alone. If the nul hypothesis proves correct within a range of high probability, the researcher accepts the result of his experiment with no further manipulation. If, on the other hand, the nul hypothesis is incorrect, new explanations and manipulations become necessary to explain the observed deviations. A probability of .05 (that is, five chances in one hundred that the nul hypothesis is incorrect) is usually accepted as the basis for further experimentation. The nul hypothesis can also be used to test the assumption that two events are related. In this case a probability of .05 again disproves the nul hypothesis, suggesting this time that the observed correlation is real. When experiments of this type are performed it is necessary to use the so-called *method of difference*, first described by the English philosopher John Stuart Mill. This method demands two types of correlation. The first states that when under controlled conditions one variable is present, the other will also be present. The second states that when, under the same controlled conditions, one variable is absent, the other will also be absent. If the correlations between variables is 100 per cent (that is, when A, then B, 100 per cent of the time; and when not A, then not B, 100 per cent of the time) no test of significance is necessary. Chance factors can upset the correlation, however, so that neither test yields 100 per cent results. The nul hypothesis is then tested to see if the distribution of the events A and B is due to chance or to a real association between them. The lower the probability that the nul hypothesis is correct, the stronger the validity of the association between variables.

Actually the two uses of the nul hypothesis described here are the same. In the first case we are testing the assumption that our model is related to a stated set of naturally occurring events and that no other events exist which disturb this relationship. In the second case we are testing the assumption that two variables are related. The first case can be stated as follows: Event A (our abstract model) is a fair representation of event B (our natural model) and no event X exists which disturbs this representation. A confirmation of the nul hypothesis in this case suggests that no event X occurs to upset the predicted outcome, and that observed variations are probably due to chance alone. In the second case, in which an attempt is made to confirm a relationship between two variables, a *rejection* of the nul hypothesis tends to prove the relationship. A *confirmation* of the nul hypothesis, on the other hand, would suggest that any observed relationship between the two variables is accidental.

It is important to note that I have used the words "tends to prove" or "would suggest." This is a correct way of stating a problem based on probability. This is true because the possibility always remains that apparent correlation is due to chance alone. The higher the probability that the relationship is not due to chance, the better the evidence that a real relationship exists. Conversely, what appear to be chance relationships may actually be real relationships.

Before we consider the structure of the basic model in population genetics, it might be well to look a bit further into some simple problems of probability statistics.

If we have a pot filled with one white bead and ninety-nine black beads, what is the probability of pulling out the white bead in a single try? The answer is one out of one hundred. Conversely, the chances of pulling out a black bead are ninety-nine out of one hundred.

Let us now alter the problem somewhat by changing the ratio of black to white beads. If we have a pot filled with twenty-five white and seventy-five black beads, what is the probability of removing a white bead in a single try? The answer, of course, is now twenty-five out of one hundred or $\frac{1}{4}$.

It should be clear from these simple examples that the probability of pulling a particular bead is a statement of the ratio between the quantity of the beads to be chosen and the total sample. The number of specific objects in relation to the total sample is known as the *frequency*. The frequency and the probability of success in pulling the correct bead in one try are identical, because the probability in this case is a function of the frequency. If the drawn bead is replaced after each try, the probability of pulling a correct bead in a new try will be the same. As long as the universe (in this case, the pot of beads) remains the same, each draw is considered as an independent event. If, on the other hand, the sample is reduced by placing each drawn bead aside, the probability will change in terms of the new total sample size and the change in frequency of the object to be chosen. If we begin with twenty-five white beads and seventy-five black beads, draw out one white bead and place it aside, the probability of drawing a white bead is then changed to twenty-four out of ninety-nine.

The problem can now be complicated a bit by increasing the complexity of the independent event. What will be the probability of pulling two white beads in succession from a sample containing twenty-five white and seventy-five black beads? The answer will be the probability of pulling a single bead once times the probability of pulling a single bead once. In this case, this is $\frac{1}{4}$ times $\frac{1}{4}$ or $\frac{1}{16}$, providing each time a bead is taken from the universe it is returned to it for the next trial.

The probability of pulling four white beads in succession will be the product of $\frac{1}{4}$ times $\frac{1}{4}$ times $\frac{1}{4}$ times $\frac{1}{4}$, or $\frac{1}{256}$. ($\frac{1}{4} \times \frac{1}{4} = \frac{1}{16}$; $\frac{1}{16} \times \frac{1}{4} = \frac{1}{64}$; $\frac{1}{64} \times \frac{1}{4} = \frac{1}{256}$.)

The probability of the same event occurring successively decreases rapidly as the expected succession is increased. It is important not to confuse the probability of successive draws as one event with the probability of a single draw as one event. The probability of drawing any object from a sample never changes no matter how many attempts have preceded any particular draw, as long as the sample size remains unchanged. A long run of good or bad luck does not change the probability of success in succeeding tries. To think otherwise is to commit the Monte Carlo fallacy!

Now let us turn to coin tossing. The probability of tossing a head or a tail in one try is $\frac{1}{2}$. Two units, one head and one tail, make up the total sample in this case. The probability of tossing two heads in a row or two tails in a row must be $\frac{1}{2}$ times $\frac{1}{2}$ or $\frac{1}{4}$. The probability of tossing one head and one tail successively is, however, $\frac{1}{2}$. This is true because there are two possible combinations available, a head followed by a tail or a tail followed by a head.

For two tosses there are three possible outcomes: two tails, two heads, or one head and one tail. The three outcomes, when added together, must equal one, because no other outcomes are possible. Now what will the expected frequencies of each outcome be? Two tails is the probability of one tail times the probability of one tail or t^2. Two heads is the probability of one head times the probability of one head or h^2. The probability of one head and one tail must be two times the probability of one head times the probability of one tail, or $2(ht)$. The total probability of the three possible outcomes can be written $t^2 + 2(ht) + h^2 = 1$.

The Hardy-Weinberg Law

The basic model in population genetics consists of a single gene locus with two alleles, A and a, which is unaffected by any disturbing influence such as mutation or natural selection. This locus occurs on a chromosome common to all members of a panmictic breeding population.

The first problem to consider in relation to this model is that of allele frequency from one generation to the next. That is, given a certain frequency in the parental generation, what will be the frequency of alleles A and a in the next or F_1 generation and in succeeding generations?

Every individual in the population will have two alleles in one of three possible combinations: homozygous dominant AA; heterozygous Aa; and homozygous recessive aa. Since each parent can donate only one allele to an offspring, parental gametes will be either A or a. The problem of gene frequency in successive generations is similar to the coin tossing problem, since the universe is made up of two variables A and a. If this is the case, A plus a must equal one, or totality. Furthermore, $1 - a = A$, and $1 - A = a$. Now, instead of one coin with a head and a tail, we have a population with particular frequencies of alleles A and a. The probabilities of the combinations AA, Aa, and aa occurring will depend upon the frequencies of the A and a genes in the population. If we let p equal the frequency of the A gene, and q equal the frequency of the a gene, then it should be obvious that the three possible recombinations of the alleles A and a from parental gametes into zygotes AA, Aa, aa should follow the formula $p^2 + 2pq + q^2 = 1$. This holds because the frequency of the two A allele combination equals the probability of the A allele times itself; the frequency of the two a allele combination equals the probability of the a allele times itself; and the frequency of the Aa combination equals the probability of the A allele times the probability of the a allele, or 2(Aa). (Remember the frequency of an event is equal to the probability of the occurrence of that event in a single-choice situation.) Thus, the frequency of A is equal to the probability (p) that A will be chosen, and the frequency of a is equal to the probability (q) that a will be chosen in a series of random combinations.

A simple problem may clarify this procedure. Let us consider a population model in which the allele A occurs with a frequency of .7, and the allele a with a frequency of .3. Then, by the formula $p^2 + 2pq + q^2 = 1$, we get:

$$.7^2 + 2(.7 \times .3) + .3^2 = 1$$
or
$$.49 + .42 + .09 = 1$$

The frequencies of the A and a alleles will be the same in every generation because the distribution and recombination of alleles follows the above formula, known as the Hardy-Weinberg Law (*Figure 35*).

While an individual can carry only two alleles for

a single trait (one on each homologous chromosome), more than two alleles for a trait can exist in a population. The ABO blood group provides an example of multiple alleles. Actually three alleles exist. Two of these, A and B, are codominant. The third allele, O, is recessive to both A and B. An individual can be type A (genotype AA or AO), type B (genotype BB or BO), type AB (genotype AB), or type O (genotype OO). The Hardy-Weinberg formula for three alleles is an expansion of $(p + q + r)^2 = 1$ or $p^2 + 2pq + 2qr + 2pr + r^2 + q^2 = 1$. The formula for four alleles would be an expansion of $(p + q + r + s)^2 = 1$, and so on.

Parental Generation

Alleles	Frequency	Numerical Frequency	Genotypes	Frequency in Population
A	p	.90	AA	p^2
a	q	.10	Aa	$2pq$
			aa	q^2
A+a	p+q	1.00	AA+Aa+aa	1

Offspring from Random Matings

Parents		Frequency of Mating Type	Frequency of Offspring		
M	F		AA	Aa	aa
AA x AA		p^4	p^4		
AA x Aa Aa x AA		$4p^3q$	$2p^3q$	$2p^3q$	
AA x aa aa x AA		$2p^2q^2$		$2p^2q^2$	
Aa x Aa		$4p^2q^2$	p^2q^2	$2p^2q^2$	p^2q^2
Aa x aa aa x Aa		$4pq^3$		$2pq^3$	$2pq^3$
aa x aa		q^4			q^4
Total		1	p^2	$2pq$	q^2

Figure 35. The Hardy-Weinberg equilibrium.

The fact that allelic recombination follows the Hardy-Weinberg formula is important for two reasons. First, it allows us to determine the frequencies of single alleles in a population if we know the frequency of the recessive phenotype, which represents, of course, the combination aa. Thus, if we collect data from a population which tells us that the phenotype aa occurs with a frequency of .09 ($a^2 = .09$), then we know that the frequency of the allele a in that population will be $\sqrt{.09}$ or .3. We also know that the frequency of A must be $1 - a$ or .7. The phenotype A can be either the genotype AA or Aa, and its total distribution in the population will be the sum of p^2 and $2pq$ which is equal to p. Where genes are codominant—alleles A and B, for example—the heterozygote phenotype will differ from either homozygote and the frequencies of either gene can be determined from the frequencies of either homozygote.

Secondly, the formula is important because it shows mathematically that the allelic frequency is constant from one generation to the next as long as random mating occurs, since random mating with the combination of two alleles follows the same rule as any similar two variable probability problem. This is important for the study of genetic change and evolution in populations because it provides a mathematical basis for determining the effects of evolutionary pressures upon populations. That is, deviations away from the constant model must be explained either in terms of chance factors or as the result of some evolutionary process.

We can now consider these pressures and the effects they have upon the genetic equilibrium of a population.

Mutation

The first change to be introduced into the population model is that of mutation. The question to be asked is: What effect will mutation have on allelic frequencies at specific loci? Again I shall begin with a simple case in which only one allele A exists at a single locus. If a mutation to a begins on that locus, and if selection is held constant, the frequencies of A and a will change from generation to generation, until such time that all the A genes will have changed to a genes. The frequencies of the alleles A and a will be different in each generation until the extinction of the A gene.

Mutations are, however, generally reversible, so that in every generation there will be some mutations from A to a and some from a back to A. The rate of mutation from a dominant to a recessive is usually higher than the back mutation. If for the sake of the model the mutations in both directions are considered as new, the allele a will build up in the population until such time as a sufficient number of genes exist which mutate back to A, establishing a new equilibrium. Thus, even in populations with constant mutation rates, the frequency of genes will eventually come to equilibrium as long as mutations occur in both directions. The new frequency of gene A in each generation before equilibrium is reached is known as $\triangle p$. If the mutation rate of A to a is called u, and the rate from a back to A v, then $\triangle p$ will be equal to $vq - up$, where vq stands for the mutation rate from a to A times the frequency of a, and up stands for the mutation rate of A to a times the frequency of A. Equilibrium will occur when $\triangle p = 0$. The new equilibrium frequency for A is known as $\hat{p}$. It is possible to solve for $\hat{p}$ through the application of the following formula:

$$\hat{p} = \frac{v}{u + v}$$

The formula for $\hat{p}$ is derived as follows:

vq = up (when vq − up = 0)
v(1 − p) = up (remember q = 1 − p)
v − vp = up
Add vp to both sides to get:
up + vp = v
Simplify to get:
p(u + v) = v
Divide both sides by u + v

$$\hat{p} = \frac{v}{u + v}$$

From the point of view of evolution, the fact that equilibriums are established between opposing mutations means that under stable conditions where selective factors are inoperant, variant alleles are maintained within the gene pool of the population. Thus the most important mechanism of variation, basic chemical changes in the hereditary material, is conserved. Environmental shifts which change the balance of selection can then operate to increase the frequency of what would become the beneficial allele.

Mutation rates in most populations are low, and these vary from locus to locus. There even appear to be genes which affect the mutation rates of other genes in carrier organisms. These low rates are themselves quite likely the product of evolution for, as I have stressed before, continuity of effective code messages is a general feature of the evolutionary process.

Selection

The effect of selection on the population model is more complicated than the effect of mutation. This is the case because selection acts on the phenotype and not directly on the genes. In a two-allele model, one must consider three possible genotypes and the degree of gene penetrance of each. If a gene is completely recessive, selective forces can act on it only when it occurs in the homozygous state. If genes are codominant, the effects of natural selection will be different for the homozygous types and the heterozygote. Dominant genes, on the other hand, will be affected equally in the homozygous and heterozygous condition. In addition to these differences, there are important cases in which a heterozygous phenotype will have a selective advantage over either type of homozygote; this condition, known as overdominance, will be discussed at the end of this chapter.

It must be re-emphasized here that most, if not all, genes are pleiotropic, or have multiple effects upon the organism. It is often the case that a gene which acts as a recessive in terms of some phenotypic marker may act as a dominant or codominant in respect to some other trait. Therefore, when discussing natural selection, it is necessary to use the terms *dominant* and *recessive* in relation to the fitness of the genes in question. A gene which has been labeled recessive, for example, because it appears to be fully masked by its dominant allele, may actually have some other effect on the organism, which gives it a selective advantage even in the homozygous condition.

With this in mind, let us look at the possible relationships between selection and dominance for two alleles at a single locus. With complete dominance and selection against the recessive allele, we should expect selection to work against the homozygous recessive only. Since fitness is a relative concept, we can consider the dominant gene as perfectly fit in relation to the recessive allele. This would give it a fitness of one. The fitness of the recessive allele would reflect the effects of selection on that gene in the homozygous condition. (Remember selection acts only on the phenotype, and the recessive trait can be expressed only in the homozygous condition.) The effects of selection would in this case be negative, since selection is against the recessive gene. Thus, the fitness of the recessive gene in the homozygous condition would have to be $1 - s$. Since selection refers to an effect on the next generation, $1 - s$ reflects a lowered fertility of the homozygous recessives in relation to the homozygous dominants and the heterozygotes, whose fitness would also be one.

The probabilities of the occurrence of the allelic combinations AA, Aa, and aa in one generation after selection would be changed by the effects of selection on aa only. Thus, instead of $p^2(AA) + 2pq(Aa) + q^2(aa) = 1$, we would get $p^2(AA) + 2pq(Aa) + q^2(1 - s)(aa) = 1 - sq^2$. The $1 -$

sq^2 term in the equation reflects the effect of selection on the total universe in relation to the previous generation in which no selection occurred. The new gene frequency of recessive gene a ($q1$) after one generation of selection is:

$$q1 = \frac{q^2(1 - s) + pq}{1 - sq^2}$$

The $q^2(1 - s)$ refers to the effects of selection on the homozygous recessive; the pq term is added to give the frequency of the recessive in the heterozygote; and the denominator $1 - sq^2$ reflects the effect of selection on the universe, which is changed by the deletion of recessive genes to selection at the rate $1 - s$. The change of gene frequency between two generations ($\triangle q$) is equal to $q1 - q$, or

$\frac{q^2(1 - s) + pq}{1 - sq^2} - q$, which is equal to $\triangle q = -$ $\frac{sq^2(1 - q)}{1 - sq^2}$

Selection can operate against the dominant. In this case, if there is complete dominance, the fitness of the dominant homozygote and the heterozygote would be $1 - s$. The fitness of the recessive homozygote is now 1. The simplified formula for the change in gene frequency due to selection is now $+ \frac{sq^2(1 - q)}{1 - s(1 - q^2)}$. In a situation where there is no dominance (codominance) and selection against the B gene, the fitness of the heterozygote will be intermediate between the fitness of the A homozygote (1) and the B homozygote ($1 - s$). Thus we would have an expectation of the following probabilities: $p^2 + 1 - \frac{1}{2}s(pq) + 1 - sq^2 = 1 - sq$. In this case, selection affects the expected frequencies of both the heterozygote and the homozygous B gene. The universe is also affected by selection on these two allelic combinations.

Overdominance and Polymorphism

In most cases selection favors one allele at a locus at the expense of the others. If selection has been operating on a locus for some time, the geneticist expects to find the most fit allele in high frequency and the less fit allele in low frequency. There are situations, however, in which two or more alleles occur in high frequency. At first it might seem as if selection were not operating in these instances. This phenomenon may be due, however, to a condition of selection in which the heterozygote is favored over either homozygote. Such a situation leads not only to a high frequency of heterozygotes in the population, but to some sort of balance between the dominant and recessive alleles, since both must be present in high enough frequency to assure the presence of the Aa combination.

An interesting example of this phenomenon is found in human populations. The chemical structure of normal hemoglobin is controlled by the dominant gene S. A mutant gene s also occurs. This gene produces the disease known as sickle-cell anemia, which is generally fatal in the early years of adulthood. The gene s has a low fitness, since homozygotes will have the disease and produce fewer offspring on the average than normal individuals. Heterozygotes for this condition normally have a somewhat lowered fitness as well, since their red blood cells have some difficulty in transporting oxygen, but their fitness is much more like that of the homozygous dominants.

Some years ago it was noticed that the gene s occurs in rather high frequencies in parts of West Africa. This was surprising since the disease is just as severe in Africa as it is elsewhere. Later it was discovered that individuals heterozygous for this condition (Ss) were resistant to a severe form of malaria found in the same regions as the high (s) frequency (*Figures 36 and 37*). Their resistance is probably related to the effects of the sickle-cell gene

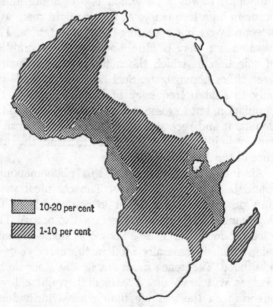

10-20 per cent

1-10 per cent

Figure 36. Frequency of the sickle-cell gene in natives of Africa.

on the red blood cell, since the malarial parasite spends a good part of its life cycle in these cells. In regions where this type of malaria is common, heterozygotes have an advantage over both types of homozygote. The homozygote recessives get sickle-cell anemia and die; the homozygote normals have a tendency to get malaria and many of them die; but the heterozygotes are resistant to malaria and have only a mild form of sickle-cell anemia. In this case we must consider the fitness of the heterozygote as one and the fitness of the other combinations as $1 - s_1$ and $1 - s_2$. We employ the sub-

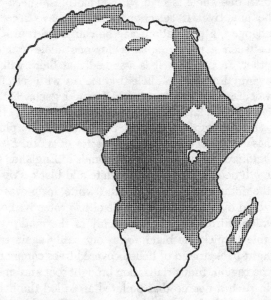

Figure 37. Distribution of falciparum malaria in the tropics of Africa. (After Allison)

scripts 1 and 2 to s since different selective forces act on the two types of homozygote. The formula for the change in gene frequency ($\triangle q$) for this situation is:

$$\triangle q = \frac{pq(s_1p - s_2q)}{1 - s_1p^2 - s_2q^2}$$

Overdominance is one case of a condition known as *adaptive polymorphism*. The term *polymorphism* refers to the existence in the same population of more than one form in high frequency. The term *adaptive* here means that the polymorphism exists due to some kind of selection in which more than one allele is favored. Polymorphism need not be adaptive. If selection pressures have recently changed, a population may be caught in transition between a high frequency for one allele and another. In this case it is assumed that there will be eventual change in allele frequencies with selection weeding out the previously favored gene. Such a condition is known as *transient polymorphism*.

An interesting case of transient polymorphism has been documented in a species of moth, *Biston betularia*, common in the Midlands of England. The moth occurs in two forms, white and black. Prior to the industrial revolution, the white form was favored over the black because of its protective coloration. Birds, the natural enemy of the moth, were able to pick the black forms out easily against the light background of lichen-covered trees common in the region. Industrialization brought soot and smoke in such a degree that polluted air killed the lichen and left the bark a dark sooty color. The white moth then stood out clearly against the dirty bark while the previous black targets became difficult to spot. The frequencies of the genes controlling color in this species began to change under the new conditions of natural selection. At the present time white moths occur in low frequency and black moths have become the common form.

Adaptive polymorphism need not occur only when the heterozygote is favored. It is also possible for selection to operate on alternate forms at different times.

The amount of genetic plasticity in a population may be due to the severity of the environment. Since this may itself vary through time, it is possible to observe populations which show a great deal of genetic variation during the course of one observation and little during the course of another. The Australian rabbit population, for example, is larger and probably much more variable just prior to the outbreak of a myxomatosis epidemic than it is at the end of such an event. Lemming populations are certainly more variable before the period of migrations and death, which occur sporadically. When disasters to large segments of any population are dependent upon such factors as density or some other cyclical event, genetic plasticity can vary with genetic rigidity regularly through time. In such circumstances the population is large and variation high when natural selection is relaxed. Conversely, when selection becomes severe, the size and the variability of the population drop. Such events tend to provide species with periods of maximal variation at one point and maximal continuity at another. Cycles of this type may lead to new evolutionary directions for a species as new variants appear in times of relaxed selection. These may then become adaptive forms when selective pressures are once more severe.

There are instances in which the cycles are so regular that two or more variants tend to exist as adaptive polymorphisms. One form is well adapted to one condition and the other form to the other. If, for example, a polymorphic species has forms well adapted to winter and those well adapted for summer conditions, then one should expect to find the summer type infrequently at the beginning of the season and common at the end. Conversely, the winter form should be numerous as the winter ends and rarer at its beginning. Such a type of variation tends to maintain the continuity of the species through time. Variation of this type occurs among wild populations of *Drosophila pseudoobscura,* a species of fruit fly.

Mutation and Natural Selection

To return to our model, it must be pointed out that mutation pressures can operate against natural selection. It is well known that certain lethal diseases, even those caused by dominant genes, continue to occur in species even though the effect of selection on them is complete. That is, every time the gene occurs, the affected individual is eliminated before it can reproduce and pass the gene on to the next generation. Whenever we have two opposing forces acting on a population we can expect an eventual balance to develop between the forces, so that the population can be said to be in equilibrium in respect to the frequency of particular alleles.

Again, the kind of balance achieved depends upon the type of allele operating. Selection acts directly on the dominant in every case, in either the homozygous or the heterozygous condition, while the harmful recessive gene is not removed from the population except when it occurs in the homozygous combination.

We may first consider mutation vs. selection on a fully recessive gene a, with the frequency q. The mutation rate to it is u, and the mutation rate from it to the dominant A is v. Selection against it is s. Note that all the terms used here have already been introduced in the sections on mutation and selection. All that remains is to combine our formulae in such a way that the balance of forces is reflected by the algebra. Equilibrium in this case will occur when:

$$up - vq = \frac{sq^2(1 - q)}{1 - sq^2}$$

In most cases the denominator and the rate from a to A will be so small as to be unnecessary for average computations. The formula can then be simplified to: $up = sq^2(1 - q)$. We may also eliminate p because $p = 1 - q$, so we get $u(1 - q) = sq^2(1 - q)$. This is equal to $u = sq^2$. We can then solve for q:

$$q = \sqrt{\frac{u}{s}}$$

This gives us the frequency of allele a when mutation and selection are in balance.

For genes A and B with no dominance, the simplified formula is:

$$u = sq \quad \text{and} \quad q = \frac{u}{s}$$

Here, sq refers to the effect of selection on all B genes wherever they occur. Selection will operate on both the homozygotes BB and the heterozygote AB, because the B gene is not masked. Thus we must include all the B genes in the formula; those which occur in the q^2 frequency BB, and those which occur in the 2pq frequency AB.

In the case of a completely dominant gene, balance between mutation and selection occurs when:

$$vq = sq^2(1 - q) \quad \text{and} \quad q(1 - q) = \frac{v}{s}$$

If H is the frequency of the heterozygotes, and if the mutant gene is rare, H would be very near the frequency of the mutant phenotype in the population. Thus:

$$H = \frac{2v}{s}$$

Genetic Drift, Interbreeding, and Inbreeding

There are three other phenomena which can have a direct effect on the frequencies or distributions of genes in a population. These are *genetic drift, interbreeding,* and *inbreeding.*

Genetic drift is merely the result of sampling error. This may occur when a small segment of a population migrates away from a parent stock. If the migrating group constituted a random sample of the original population, one would expect the frequencies of genes in the new population to be a fair representation of the old. If, however, the migrant population were small, then it would be likely that only a portion of the original gene pool would be represented. Even if all the genes were represented, however, they would probably occur at frequencies different from those of the original population. Such a change would be due to a non-random sample of genes rather than to selection or mutation. When genetic drift occurs, it is possible for a gene to be lost merely because it is not represented in the sample, or for the frequency of a rare gene to rise significantly. The latter occurs when a migrant gene begins at a higher frequency in the new population than in the old. This, again, is merely the result of accidental changes in gene frequency. It has been known for some time that small island populations which have developed from a common parent stock are often widely divergent in genetic structure. This phenomenon, known as the *founder effect*, is the result of genetic drift. A small number of individuals serve as the "founders" of future generations, and their genes lie at the base of the new gene pool which reaches equilibrium after a few generations. The existence of genetic divergence between isolate populations cannot be taken as an indication of natural selection, since genetic drift may have had a more profound effect on those populations than the much slower process of selection. But, by the same token, the reverse situation of wide-ranging similarities between such populations should not be accepted as a result of natural selection; however, one might argue that such similarities are due to the effect of natural selection which has overcome the effects of drift. In such cases similarities between dispersed groups might well be due to the effect of gene flow acting to reduce genetic divergence. The reason why selection is often rejected as an explanatory hypothesis in the latter case is the fact that natural selection is opportunistic, i.e. it operates on what already exists in a population. If two populations differ genetically from each other because of genetic drift, natural selection will go on from there working on those genes which are already present. Under these conditions it is unlikely that selection would ever bring the two populations back to identity. A divergence, once established, is likely to continue. The founder effect then can lead to a situation in which the original differences between populations which were derived from a single stock are due at first to genetic drift, and then to selection working in different ways on the two different existing gene pools.

Genetic drift may also occur without migration. Small populations in general are subject to nonrandom accidental changes in gene frequency which may have little to do with either selection or mutation. Natural disaster can wipe out a significant percentage of such a population without regard for selective coefficients and thus change the constitution of the gene pool. In human populations this factor can be even more important than in other species. Social regulations which affect the potential reproductive capacities of individuals may interfere with their actual biological fitness. Thus, a wealthy man in a polygamous society may have many wives and children while a poor man may produce few offspring. Celibacy imposed on a segment of the group through such beliefs as religious convictions may reduce the fertility of potentially fit individuals to zero. The fact that human groups were small throughout most of man's history, coupled with the fact that social rules differ from group to group, has certainly served to widen the genetic gaps between populations. This divergence has been reduced, however, by what must have been and continues to be a high degree of gene flow between potentially divergent populations. Man has always been a highly mobile species, adaptable to a variety of environmental conditions. It is unlikely that true isolates ever existed except in highly unusual circumstances, and then for relatively short periods. Remember, selection is an extremely slow process, and man is an extremely young species.

Man is also the only animal in which breeding is regulated through the operation of social rules. Incest taboos of one sort or another exist in all human groups. This means that panmixia is never reached in human populations, since mother-son, father-daughter, and brother-sister matings only very rarely occur. On the other hand, genes are dispersed widely through most human populations, and the regulations against close marriages tend to keep a relatively even flow of genes moving through the segments of the group. Close inbreeding in marriage for any period of time would tend to increase homozygous loci, since rare genes occurring within a family would be more likely to pair up. Cousins, for example, might share the same lethal alleles by virtue of inheritance from common ancestors. If such cousins were heterozygous for a particular condition, neither one would be affected. If, however, they married, and if the disease was based on a single-gene effect, there would be a 25 per cent possibility that some of their offspring would be affected. An outmarriage for such individuals would be less likely to result in homozygosity, since their mates would be less likely to carry the same lethal genes. It has been estimated that all humans carry at least three lethal alleles at different loci. The frequency of genetic disease is lowered by the fact that given the total number of loci carried by humans, the chances of homozygosity at any lethal loci is quite low. Inmarriage increases the chances that such lethal genes will be unmasked. Regular patterns of inbreeding would tend to change the frequencies of certain genes by unmasking them, thus making them subject to natural selection. Inbreeding also upsets expectations based on the Hardy-Weinberg law, since such combinations are nonrandom.

Interbreeding between populations which differ markedly in genetic constitution presents another case of modification. A cross between two such groups creates a third and different group which presents a different face to the process of natural selection. Interbreeding brings about changes in gene frequency and type just as much as mutation and natural selection, and, of course, the process is much more rapid.

A problem is created by the occurrence of what might be gene flow, genetic drift, or natural selection in a series of lineally distributed populations. If an investigator discovers differences between isolate or semi-isolate populations, how is he to determine what evolutionary forces have been acting on them? I have already suggested that similarities between isolates strongly support the hypothesis of gene flow, but in a linear sequence one could expect a gradual diminution of the frequency of a trait. How might one separate the effects of gene flow, drift, or natural selection from one another? In theory the answer is fairly simple. Genetic drift should lead to an unordered distribution of gene frequencies. On the other hand, if natural selection has been operating on a series of relatively contiguous populations in such a way as to produce a change in gene frequency, one should expect the genetic variation to occur coterminously with any environmental gradation. Such a congruence between a clinal distribution of a trait and shifts in the environment are strong evidence for natural selection. Finally, if environmental shifts are not geographically coterminous with a cline, then gene flow is the better hypothesis. Unfortunately, this procedure is not always as easy as it sounds. In the first place, the relationship between some aspect of the environment and the gene in question must be found, and in the second place, populations may not remain conveniently in place long enough for the cline to be as neatly congruent as it should.

It has been suggested, for example, that skin color in man is correlated with the amount of exposure to solar radiation. According to the theory, an environmental gradation exists from the polar region running south with a constant increase in sunlight as one approaches the equator. The cline in question is the general gradation of skin color, which is largely, but not completely, lighter in the northern regions and which becomes progressively darker toward the equator. There are, however, several problems with this hypothesis. In the first place, the distribution of ultraviolet radiation may be the crucial selective force (in association with skin-cancer incidence) and the relation between sunlight in the visible spectrum and ultraviolet is not coterminous. Thus, the apparent environmental gradation may not be as neat as it looks. In the second place, as I have already stated, human populations are highly mobile, and there are certainly populations which no longer fit the pattern. Historically we know that Negroes are recent to the New World and that their distribution in the North of the United States and Canada can be explained away as a recent phenomenon. We know less, however, about the populations of Southeast Asia, and it is difficult to say with conviction that the gradation which now exists has been placed geographically long enough for natural selection to have produced it. In the third place, some scientists have questioned the role of skin color alone in the etiology of skin cancer. It is possible that some other factors related to the thickness of epithelial tissue is the crucial factor, and that this trait may not follow the distribution of skin color.

This does not mean that the hypothesis is incorrect. More research is necessary to confirm it. These problems do, however, point up the difficulties which exist in the study of population genetics and evolution.

V
EVOLUTIONARY SYSTEMS

In the preceding four chapters I have examined the genetic basis of evolution. I have tried to show that the key to the entire process rests in two sets of opposing mechanisms. One of these sets operates to maintain stability; the other provides the variation necessary for further adaptation. Evolution is an adjustive process in which species respond to continuous environmental pressures. In one sense the response is passive, since it depends upon existing random variation; but in another sense it is directional, since adaptive trends (or patterns of fit) develop in relation to particular environmental settings.

Two Meanings of Adaptation

It is necessary at this point to clarify the term *adaptation*. It can, and indeed must, be defined in two ways. First it can be seen as a temporal process of positive feedback in which transgenerational changes are directed by selective forces in the environment. This in sum is evolution. Adaptation also refers to response processes which occur within individual organisms as they adjust through negative feedback to environmental changes during their own lifetimes. All organisms are dynamic systems composed of parts which together form subsystems, each of which plays some role in the maintenance of life. These subsystems are geared to minute changes which constantly occur within the living space of the individual. Homeothermic, or warm-blooded, animals, for example, must constantly adjust their chemical and physiological reactions to variations in external temperature. These adjustments are adaptive for the ongoing organism-system in the second meaning of the term, and they continue as long as life continues. Animals and plants can also adapt this way to long-term environmental change such as colder temperatures or drier conditions. Such adaptations have been observed in men who have been exposed to cold or heat stress and high altitudes. Individuals from a temperate climate who are subjected to extreme cold will modify their bodily reactions sufficiently to overcome moderate cold stress. The same individuals placed in an experimental climatic chamber, which is capable of simulating tropical heat and humidity, will develop a considerable tolerance to heat stress. In such cases the amount of work which can be done without noticeable fatigue will increase daily until a measurable limit is reached. This kind of adaptation is a reflection of the inherent flexibility of organisms. It is as much a product of evolution as are transformations which occur through time as species evolve. The adaptive changes which occur in response to stress are not inherited (these are acquired characteristics). It is the ability of organisms to tolerate variation and to adjust to it that is most certainly a product of evolution. What has evolved in these cases is an incredibly large and complex set of feedback mechanisms, similar to those discussed in Chapter III. In a very real sense, all living things can be thought of as delicate biological systems which maintain their integrity through the operation of adjustive mechanisms which constantly operate in response to environmental pressures.

Adaptation to environmental variation by an organism in its own lifetime is often referred to as *physiological adaptation.* Physiological adaptation provides flexibility and therefore allows populations to respond to short-term variations in environmental conditions without genetic change. Selection is likely to favor the development of such flexibility because stress conditions will eliminate those organisms that are unable to adapt physiologically. As far as evolution is concerned, the development of physiological adaptation can be classified as a mechanism of continuity, since flexibility of response leads to genetic stability over the course of generations. In other words, physiological adaptation inhibits transgenerational genetic change because it buffers particular genotypes from the effects of moderate environmental change. Such an effect, however, must be seen as an artifact, or incidental benefit, of selection because the actual selection process in this case acts to preserve those organisms in a population that are capable of surviving under variable conditions.

Different types of physiological adaptation to the same kind of stress may occur depending upon the degree of intensity and duration of the stressor. Thus, in response to cold, humans will first shiver. Continuing cold will lead to the stimulation of hormonal secretions that will act to increase peripheral blood flow and metabolic activity. Physiological

adaptations that come into operation in orderly
fashion depending upon the intensity and duration
of stress can be fit into a *hierarchy of responses*.

Transgenerational or evolutionary adaptation can
be considered in terms of two adaptational modes.
These have been referred to by A. R. N. Radcliffe-
Brown, an anthropologist, and Morris Goodman, a
biologist, as internal and external adaptation. To
quote Radcliffe-Brown, "A living organism exists
and continues to exist only if it is both internally
and externally adapted. The internal adaptation depends on the adjustment of the various organs and
their activities, so that the various physiological
processes constitute a continuing functioning system by which the life of the organism is maintained.
The external adaptation is that of the organism to
the environment within which it lives. The distinction of external and internal adaptation is
merely a way of distinguishing two aspects of the
adaptational system which is the same for organisms
of a single species."

Biological Systems

Any organism can be thought of as a single grand
system, but it is often useful to focus attention on
one or more subsystems which can be analyzed
separately in terms of environmental relationships.
When the geneticist speaks of the adaptive value
of a single trait, he is fully aware of the fact that
he must not only consider this trait in relation to a
specific environment, but that it is also related in
an almost infinite variety of ways to other traits of
the whole organism, and that these traits often are

parts of systems which in turn are related to other perhaps more complicated systems.

Biological systems of varying complexity make up the subject matter of biology and are a major concern of students of evolution. While each system is in many ways unique, it is the task of the scientist to seek out uniformities which are common to all the phenomena under consideration. In evolutionary biology one important analytic task is to sort out the similarities which exist among various levels of organization from the subcellular to the population, and even beyond to the *ecosystem*. (The latter is composed of all the species living in a particular environmental zone which in some way interact with one another, and all those physical aspects of the zone which affect one or more of the species.)

Biological systems are characterized by a series of attributes, the most important of which is self-regulation. Other attributes of general application are:

1. *Subsystems*

Biological systems are composed of units which may themselves be systems. The lowest-order system is composed of units which are not themselves biological systems. Thus cells which are biological systems contain molecular units which are not biological systems. The simplest complete biological system is the single-celled organism. Other levels in order of complexity include: the metazoa, populations, and ecosystems.

The concept of populations and ecosystems as systems must be approached with caution. It must be stressed that population genetics does not describe the behavior of a system, but rather makes predictions about average outcomes affecting individual members of some universe. Each member of a population, however, exerts some pressure on each other member, and hence the entire population can be thought of as a system. This holds true for ecosystems as well, since constituent populations are linked into an adaptive network. The selective value of a gene may depend upon the characteristics of the population in which that gene occurs or on other linked populations. Changes in one segment of a population or ecosystem may produce regular and predictable changes in another segment.

2. Self-replication

Biological systems are temporally finite and must therefore maintain themselves transgenerationally through some process of replication. Complex biological systems such as populations and ecosystems are replicated through the reproduction of constituent units. It is the reproductive process involving individual organisms which provides the basis of both continuity and variation.

3. Creativity

Biological systems have a tendency to change transgenerationally in the direction of maximized efficiency. Maximized efficiency can be defined in terms of the self-regulatory capacities of the system. Some systems are more efficient than others in the maintenance of homeostasis under environmental pressures. These can compete successfully with other less efficient systems. Maximization can be measured in terms of population size in regard to an environmental niche. Reproductive efficiency has always been a standard measure of evolution, and reproductive success is a function of adaptation. It makes little sense, however, to compare the numbers of one species to the numbers of another if the two species are not engaged in competition. Population size as a measure of maximization is useful only within the context of a particular environmental niche. A single niche can be exploited successfully only by a single species. Thus, if interspecific competition for a niche occurs, differential reproduction becomes a measure of successful exploitation. A successful competing species will replace a less successful one. Marsupials, for example, a group of what are now relatively insignificant animals, were once widely distributed and successful. The development of more efficient placental mammals led to the replacement of marsupials, except in a few ecological zones where they faced no substantial competition. Marsupials are now restricted, with few exceptions, to the continent of Australia, which became an isolated land mass before the evolution of true placental mammals. When a niche is controlled by a single species, the reproductive success of various individuals and, in some cases perhaps populations, may reflect intraspecific competition and maximization.

Since gene flow is normally considerable between adjacent or overlapping populations of the same species, interbreeding would result in increased variation in a newly combined gene pool with selection tending to favor the most fit individuals. Population competition and selection, as opposed to individual selection, can occur only when such groups are organized into social units which for some behavioral reason (either genetic or nongenetic) do not interbreed. The competition is then between whole social units. Human groups often fit this model, and one can, in some circumstances, speak of population competition and group selection in man, although we must never forget that interbreeding does occur frequently between human groups.

Species which are able to maximize their self-regulatory capacities in relationship to more generalized environments may widen their niches and override previously successful species. Warm-blooded animals, for example, spread into a wide range of environments differentiated by temperature competed successfully with less well-adapted cold-blooded forms. The niche-widening capacity of a species or population is also a reflection of maximization, but it, too, can ultimately be measured in terms of reproductive success. A higher degree of maximization means that more individuals can survive and exploit an environment.

When change occurs transgenerationally, a transformation from one system to another can be said to have occurred. Each new system is dependent to some degree on the system directly preceding it. The tendency for systems to become transformed in the direction of maximized efficiency is not a teleological process. Variations occur with high enough frequency so that there is a substantial probability that at least some will be creative in nature.

4. *Specificity of adaptation*

The tendency for maximized efficiency or adaptation is always relative to a specific environment. There is no such thing as absolute adaptation. The reason why general trends can be sorted out of evolutionary history (the development of increasing complexity in the central nervous system is one good example) is that these trends reflect increasing adaptive capacities in relation to a wide range of environments. An improvement in self-regulation can have wide applicability to environmental variations and therefore remain established in many, *but not necessarily all,* succeeding forms. So-called regressions in evolution are not violations of some general rule of "progress" but represent specific adjustments to environmental requirements. For the same reason, the belief held by some that evolution represents the general development of progressive complexity is erroneous. Again, while the sequential development of complexity of the nervous system is a fact of evolution, the reduction in the number of bones of the vertebrate skull, and therefore simplification, is also a fact of evolutionary history. In both cases these developments are related to specific, if widely distributed, encounters between biosystems and specific environments, and to the fact that systems are dependent to some extent on the directly preceding systems.

5. *Redundancy*

On the ecosystem level, and on other levels as well, the development of alternative pathways to self-regulation has an obvious selective advantage. In a food cycle, for example, if alternate species are

available, the loss of one link in the chain would not
threaten the over-all maintenance of the system. If,
on the other hand, a food chain was dependent on
a single species and it was eliminated, then the
whole system would be liable to collapse. On the
individual level, the dual functions of some organs
of the body and their ability to take over the operations of some damaged parts of the system serve to
protect the organism from serious breakdowns. Redundancy in systems is an adaptive device, since
it provides alternative pathways for survival. This
development can account for some of the complexity which exists in many *but not all systems*.

Such complexity, where it exists, can have another important effect on the development of maximization in systems. These redundant parts provide
a certain stability for the system under stress, but
at the same time a greater differentiation of parts
has the added advantage of storing potential variation which can be exploited in different ways if the
environmental situation is changed in some way.
Thus we have another case in which both continuity
and variation are maintained within a single phenomenon.

6. *Distribution*

Biological systems are distributed both spatially
and temporally. Spatial distributions represent synchronic maximization in reference to specific environments. These distributions can be studied by
population geneticists and physiologists in attempts
to discover what mechanisms are present in populations which maintain their efficiency. Temporal
distributions represent sequential changes in sys-

tems. These can be studied by the paleontologists in attempts to discover the histories of specific system transformations as well as the relationships between these and environmental changes over long periods of time.

7. Characteristics of ecosystems

Ecosystems may be classified as *generalized* and *specialized*. In generalized ecosystems, there is an abundance of species each of which is represented by a relatively small number of individuals. Specialized ecosystems are characterized by a small number of species each of which has a relatively large number of representatives. Under natural conditions, the most generalized ecosystems are found in the tropical forest and the most specialized in deserts and on the arctic tundra. In the tropics most of the available energy is stored in growing plant forms. Destruction of the forest cover removes available energy from the system and degrades it. Agriculture in such an environment, which does not involve short growing periods (one or two years of cropping) and long fallow periods, or fertilization, may rapidly destroy soil productivity.

In artificial ecosystems affected by man, primitive patterns of agriculture that involve numerous species, and all other types of polycropping, are generalized. Monocropping, so characteristic of modern farming, produces a highly specialized system.

In both natural and artificial ecosystems, the generalized type has the advantage of redundancy. In addition, attacks by insect pests and soil depletion are less likely to occur where there is a diversity of species.

All ecosystems contain food chains in which species are linked through a process of energy exchange. Plant foods lie at the base of these food chains. Next in order are herbivores, then carnivores and secondary carnivores. Organisms such as bacteria that degrade dead animal and plant material to organic compounds are also linked into food chains. At each level of a food chain, energy is lost to the system through entropy. According to Margalef (*Perspectives in Ecological Theory*, 1968), if one were to draw a chart for every ecosystem, it would be possible to deduce the relative amount of total energy loss produced in the maintenance of the system. If two systems were similar, the one with more species and more interactions between them would be associated with weaker exchanges of energy and a smaller destruction of information (less entropy). In this type of system, the same energy loss drives a more complex system, containing more information. The cost (in energy) of such a system is lower than in a specialized system.

Natural ecosystems are subject to a process of *succession* in which species replace each other through time. The final assemblage of living forms in a process of succession is known as the *climax*. No climax is definitive however. Such rapid natural processes as fire and flood and such slow processes as climatic and geological change reduce natural ecosystems to more immature types. Evolutionary change can also alter the sequence of succession and the climax.

According to Margalef, natural selection is not wasteful because it uses energy that would be lost in keeping an ecosystem running. In terms of succession: "The energy gates at the places where species interact—or where they interact with environment—are the organs by which selection is achieved and evolution occurs, the rate of evolution depending on the efficiency of the gate. Succession in any self-organizing system involves the substitution of some piece of the system by some other piece that allows the preservation of the same amount of information at a lower cost, or the preservation of more information for the same cost. In this context, information is anything that can influence and shape the future, and the cost is represented by the energy used, which amounts, practically, to the energy entering the ecosystem as primary production.

"Evolution cannot be understood except in the frame of ecosystems. By the natural process of succession, which is inherent in every ecosystem, the evolution of species is pushed—or sucked—in the direction taken by succession, in what has been called increasing maturity."

Molecular Behavior

In addition to the characteristics outlined thus far, many, but by no means all, biological systems on the organism level are capable of whole-system responses to environmental stimuli. This means that the organism as a unit is capable of integrated activity, or what might be called *molecular behavior*. Coordinated movement, which is a feature of most animal and some simple plant species, is a prerequisite for such behavior. This type of response pattern differs in complexity and form from species to species, but in general it represents a powerful line of adaptation, since it enables organisms to change their environmental field by changing their own position within that field. The development of many somatic traits shows correlations with various types of molecular behavior, and there is no doubt that evolutionary forces act on organisms in terms of behavior as well as morphology.

The interrelationships which exist between organismic form and function, and the existence of molecular behavior in many species, have led many researchers toward an analysis of biological systems in terms of mechanical models. One of the most interesting of these is Ashby's *Design for a Brain*. Noting the similarities between feedback systems in machines and in organisms, Ashby constructs a logical model for the operation of self-regulating biological systems which are capable of responding to environmental stimuli. Feedback systems built into this model are capable of making adjustments to information which is fed into the system through preceptors. Information which accumulates through encounters with different environmental stimuli is stored so that more and more appropriate responses can be made as the system accumulates experience. In the following chapters I shall attempt to outline the evolutionary basis for the development of such systems and to account for the very complex type of behavior which characterizes the human species.

VI
BEHAVIORAL GENETICS

Most books on evolution, even those which deal with evolutionary mechanics, emphasize the development of somatic traits. These are then discussed as adaptive units which can be measured in terms of selective coefficients. This method of explanation is derived from experiments and field observations which, for the sake of simplicity, are designed to test hypotheses linking traits to environmental parameters. The traits chosen are usually unambiguous units which can be catalogued in terms of specific physical characteristics. Nonetheless, extrapolation from these experiments to theory represents a somewhat artificial approach, since no trait has an adaptive value except in reference to some other traits of the system under discussion. The problem is handled statistically. Adaptation is measured in terms of the *mean survival value* of a gene: differences among organisms carrying the gene are "randomized out"—that is, the selection coefficient reflects the average value of the gene on all backgrounds.

Of equal importance is the fact that there is a wide variety of genetically controlled behavioral traits which have no easily apparent physical connection but which are crucial to the survival of organisms. Furthermore, if we remember that evolution involves the transformation of self-regulating systems in the direction of maximized efficiency within specific environments, a considerable part of the theory must deal with both the response patterns of variables and the behavior of the total system.

In this chapter I shall consider the problem of behavioral genetics, and in the next pass on to a discussion of the prerequisites of behavioral systems and their evolutionary development.

A series of recent experiments in the rather new field of behavioral genetics has clearly demonstrated that many behavioral characteristics are genetically determined and that these characteristics follow the classical genetic model presented in the earlier chapters of this book.

Genes and Simple Behavior

Professor Jerry Hirsch of the University of Illinois is one of the pioneers in this field. Working with the favorite laboratory animal of geneticists, the fruit fly *Drosophila*, Professor Hirsch demonstrated that the normal behavior of fruit flies to fly in an upward direction against the pull of gravity could be reversed by artificial selection. Hirsch constructed a maze which consisted of a series of alternate-choice points. At each choice point the insect could fly either down or up. Most of the flies in the original sample tested wound up at the top of the maze as expected, but some of them, representing variant forms, appeared in various levels below the highest point. These insects were then collected, and their offspring run again. After each run, flies which exhibited a downward choice were preserved and bred. After several generations, Hirsch's efforts resulted in a strain of downward-flying fruit flies. What he had done was to exploit a naturally occurring variation within the species. The experiment was of the same type carried out in a less rigorous way by animal breeders, who have always considered the behavior of their animals as one aspect of economical stock breeding.

After the pure strain of down-flying insects was established, Hirsch was able to cross them back to the wild strain to determine the genetic structure of the trait. As would be expected, it followed the normal Mendelian rules of inheritance. In another experiment he relaxed selection pressures on the flies to see if succeeding populations would return gradually to the more normal pattern of behavior. This is indeed what happened, and this fact suggests that natural selection (as opposed to Hirsch's artificial selection) is what produced the original pattern in wild flies.

Genetically controlled behavioral variation has also been demonstrated in a number of other species. One of these is the laboratory rat.

Several years ago Professor R. C. Tryon of the University of California at Berkeley decided to test the accepted dictum that individual rats were psychologically equivalent—that is, that one rat is just about the same as any other rat in terms of behavior. Again, Tryon was interested in exploiting naturally occurring variation in rat populations. He constructed a rather complex maze and artificially selected two groups of rats: those which performed well in the maze and those which performed badly. "Intelligent" rats, i.e. those which did well, were inbred and eventually a strain of these animals (known as the S_1 strain) was developed. A strain of poor performers (S_3 rats) was also isolated through artificial selection procedures. For several years it appeared as if Tryon had actually selected out strains of smart and stupid rats (relative to some standard of average rat intelligence). In the last few years, however, a group of investigators at Berkeley showed that the S_1 animals were good only at one particular kind of maze based on spatial cues. In fact, the S_3 rats did better than the S_1's in certain other maze experiments. This discovery suggests strongly that intelligence is actually a complex of traits associated with behavior, and that more specific and carefully defined units must be tested in selection experiments.

An analysis of specific behavior underlying the differential responses of S_1 and S_3 rats revealed the following: In three out of five maze measures, in performance dulls were either equal to or better than the brights. According to the full analysis, brights were more food driven, low in motivation to escape water, timid in open-field situations, more purposive, and less destructive. Dulls were not food driven, they were better or average in motivation to escape water, and they were fearful of mechanical apparatus features.

Meanwhile at the Roscoe B. Jackson Laboratories in Maine, a whole series of inbred mouse strains had been developed for cancer research. Selection procedures in these animals were related only to genetic propensities for, or resistance to, the development of induced and spontaneous cancer. Pure lines were created in order to hold genetic factors constant so that attempts could be made to determine the effects of environment on phenotypic variance. Some time later, when geneticists became interested in behavior, these mice were tested to determine if strain-specific behavioral differences had accidentally been separated out along with the somatic traits originally selected. In every case it was found that pure-strain mice display strain-specific behaviors. One of the most interesting outcomes of this investigation was the discovery that one strain of mouse when presented with a choice between water or alcohol consistently chose alcohol as the preferred beverage. An intensive study of "alcoholic" mice followed, and it turned out that a single gene could be related to alcohol preference. In addition to this gene, alcoholic mice were shown to carry another gene at a separate locus which affected the metabolism of alcohol in the system. Not only did alcoholic mice like alcohol, but they were able to metabolize it efficiently. Exploiting the phenomenon of crossover in meiosis, these genes were separated in hybrid strains. New crosses produced mice which could metabolize alcohol but preferred water and mice which preferred alcohol but which could not metabolize it.

A more ambitious experiment in behavioral genetics has recently been completed at the Jackson Laboratories. There John Paul Scott and John Fuller studied five breeds of dog for twenty years to determine the genetic basis of behavior. Their first task was to determine if behavioral breed differences could be sorted out if environmental variables were held constant. The dogs studied included cocker spaniels, basenji hounds, beagles, Shetland sheep dogs, and fox terriers. After considerable study in which all dogs were raised under the same controlled conditions, Scott and Fuller were able to demonstrate that what breeders had assumed to be true was indeed a fact: different breeds of dog had characteristic behavioral traits which breed true.

In the next phase of their study, Scott and Fuller crossed cocker spaniels and basenji hounds. These breeds were chosen for two reasons. First, they were assumed to be the furthest apart genetically due to long-term isolation (the basenjis in Africa and the spaniels in Europe and America). Second, the behavioral traits exhibited by these breeds fell on opposite sides of certain behavioral dimensions. The basenjis were aggressive dogs; the cocker spaniels rather docile and affectionate. Hybrid crosses between basenjis and cockers produced independent assortment and segregation of behavioral traits just as would be expected in the case of somatic traits. Back crosses of the F_1 hybrids to pure parental strains helped the researchers determine what kind of genes were operating in the control of behavioral traits. Surprisingly many of these traits appeared to be based on the operation of single genes, although some were apparently polygenetic.

An interesting hypothesis tested by Scott and Fuller was the assumption that a specific gene or genes for tameness was a general characteristic in all dog breeds, and that if this was the case it is likely that domestication from wild species occurred only once. They reasoned that if dogs were domesticated more than once, different genes for tameness would have been selected out and that crosses between breeds would produce F_1 hybrids which would be wilder than either parent strain. This would occur because genes for wildness masked in the homozygous pure strains would be unmasked in the heterozygous hybrids. The hypothesis that dogs were domesticated once was supported by the fact that between-breed crosses do not produce differences in tamability.

Scott and Fuller also found that such characteristics as aggressiveness and timidity were controlled by different loci, and that therefore they could not be considered opposite aspects of the same behavioral dimension. This means in effect that it is possible to have a dog which is both timid and aggressive, as well as nontimid-aggressive, nonaggressive-timid, and nonaggressive-nontimid animals. A puppy cowering in the corner but snarling at anything or anyone that approaches it might be an example of a timid-aggressive animal. This finding illustrates the fact that behavioral traits are often difficult to study because the units of behavior to be analyzed are not always easy to sort out in advance. This is exactly the same problem faced by workers exploring such gross factors as intelligence, which is obviously a complex of traits. The word is merely a catchall term masking the ignorance of the researcher.

Like somatic traits, behavioral traits are only indirectly controlled by genes. Remember, genes produce enzymes that generally have multiple effects on the metabolism and hence development of the organism. Thus at the fundamental level, genetic control relates to the maintenance or, in the case of mutation, to the alteration of basic chemical processes. At the level of the emergent phenotype, genes play their role *only* in combination with environmental effects. Furthermore, close analysis will show in most cases that factors on the level of basic organismic reactivity underlie the control or modification of specific behaviors. Metabolic pathways control the reactivity of the organism to stimuli coming from its environmental field. Thus chemical changes may alter the perceptual thresholds of organisms in one or more perceptual modes such as vision or susceptibility to tactile stimulation. These thresholds may be raised or lowered, rendering the organism more or less sensitive to particular outside stimuli. In addition, the response pattern itself may be altered so that an organism might absorb more or less environmental information via its sense organs before responding to it. Finally, the actual reaction might be altered. So, for example, approach might replace avoidance. As has been demonstrated with S_1 and S_3 rats, behavior that is the end product of complex interactions may be caused by the interplay of many factors any one of which might alter the total outcome.

Most work in behavioral genetics has been performed on pure-strain organisms that have been bred in the laboratory in order to control genetic variation. Recent experiments on crosses between two pure strains of the same species show that hybrids may perform better than either parental type. Such a result may depend upon hybrid vigor, that is, the heterozygotes produced by the cross might be generally more viable than either set of pure-strain, homozygote parent. John H. Bruell, experimenting with exploratory behavior in pure-strain and hybrid mice, found that hybrid mice tended to explore more than their inbred parents. This result is significant because explorative behavior is a major aspect of adaptation in animals.

Human Behavioral Genetics

Human behavioral genetics is an extremely difficult subject to study. In addition to the problem of slow maturation and a consequently long time between generations, the fact exists that for moral reasons humans cannot be subjected to breeding experiments like fruit flies or pea plants. Also, it is almost impossible to separate environmental factors from genetic effects in behavioral studies. Those few cases in which really sound relationships have been established between genetic factors and behavior in human beings have all been related to mental deficiency of one kind or another. Both phenylketonuria and hypothyroidism referred to in Chapter I have obvious behavioral consequences, and Down's syndrome mentioned in Chapter III is also a genetically based behavioral condition.

Those studies that deal with human behavioral genetics depend primarily upon the analysis of data on identical twins reared apart and together. Since identical twins have identical genotypes, any phenotypic differences are assumed to be due to environmental effects. Identical twins reared apart constitute a natural experiment in which genetic effects are held constant and environmental effects varied. Twin studies dealing with behavior have centered primarily upon mental illness and intelligence.

Studies of intelligence within particular population groups (mostly white Americans) suggest a hereditary factor ranging from about 45 per cent to 80 per cent. But just as detailed analysis of Tryon's rats revealed complex differences between "dull" and "bright" rats, so responses to I.Q. tests might turn out to obscure a series of very complicated behavioral patterns.

The most unfortunate aspect of I.Q. studies, however, is that they have been extended with little caution to so-called racial differences. It is a well-known principle of genetics that estimates of hereditary effects in one population cannot be used to account for differences between that group and some other group, that is, unless the hereditary element produces the entire effect. This is because environmental effects on the phenotype may differ strongly from population to population. It has been estimated, for example, that the hereditary element in height is about the same for all human groups (the general estimate is about 70 per cent). Yet the differences between pre-World War II Japanese and Americans appear to be due in large measure if not totally to differences in diet. In other words, close to 100 per cent of the phenotypic differences are due to the environment even though 70 per cent of the trait is hereditary.

The recent publication by Arthur Jensen of an article entitled "How much can we boost I.Q. and scholastic achievement?" in the *Harvard Educational Review* (Vol. 39, 1969) has once more raised the question of genetics and I.Q. differences between whites and blacks in the United States. Jensen suggests that about one half the difference in performance on I.Q. tests between blacks and whites (15 points) is due to hereditary factors. As he himself admits, his method suffers from the usual extrapolation from studies of white identical twins to differences between whites and blacks. In addition, errors in statistical treatment and a weighting of data in favor of his hypothesis call his results into question. As things now stand, most geneticists agree that in human genetics the separation of hereditary and environmental factors in group comparisons of behavior is, at the very least, a dubious proposition.

When black versus white intelligence is the question, the delineation of the real populations concerned becomes important. I. I. Gottesman, in a book edited by Deutsch, Katz, and Jensen, discusses the geographic range of populations in Africa from which slaves were imported to Charleston during the period 1733 to 1807. His figures, taken from a study by William Pollitzer, show the following percentages: Senegambia, 20 per cent; Windward Coast, 23 per cent; Gold Coast, 13 per cent; Whydah-Benin-Calabar, 4 per cent; and Angola, 23 per cent. This distribution covers more than one thousand miles of coastline and extends six hundred miles inland. Thus the range of genetic and ethnic groups tapped was extensive.

Gottesman goes on to point out that to speak of a single white or black population in the United States is a tremendous oversimplification. The heterogeneity of black Americans prevents us from speaking of an "average black American" with a specific percentage of white genes. Gottesman ends his article by stating: "At the present time Negro and white differences in general intelligence in the United States appear to be primarily associated with differences in environmental advantages" (page 46).

A recent study by J. J. Sherwood and M. Nataupsky suggests that studies of differences in I.Q. between whites and blacks reveal more about the analysts concerned than they do about the hereditary or environmental factors involved in these differences. Biographical characteristics of the scientists themselves correlate with interpretations of collected data. Those who believed that intelligence differences were innate tended to be first-born. Those whose grandparents were foreign-born tended to see no differences between whites and Negroes while those whose grandparents were American by birth indicated definite findings of innate inferiority in Negroes. Those whose parents had low mean years of education saw environment as a crucial factor in test responses. Individuals from rural backgrounds were more likely than those whose origin was in urban centers to find indications of biological differences. Biological differences were also stressed by those who had higher scholastic standing as undergraduates. Those who tended to accept genetic factors in racial differences in I.Q. tended to be youngest in publication of research, first-born, with American-born grandparents, with parents who had many aggregate years of schooling, of rural backgrounds, and with high scholastic standing as undergraduates. It is apparent from these data that those researchers who found indications of Negro inferiority came from higher socioeconomic backgrounds than those who took the egalitarian position.

In the 1930s and 1940s W. H. Sheldon attempted to systematically relate physical structure and behavior. First he developed a classificatory system of physical type based upon components of fat (endomorphy), muscle (mesomorphy), and leanness (ectomorphy). Each characteristic was scored on a seven-point scale with mixed values given to intermediate types. Sheldon then rated his subjects for behavioral characteristics and sought correlations between these and each somatotype. While there is some doubt that Sheldon's system successfully isolated three separate physical dimensions (there is evidence that ectomorphs and endomorphs lie at the extremes of the same dimension), his was a pioneer attempt to find physical markers for particular behavioral types. This work is a kind of crude behavioral genetics in which neither genetic nor environmental data is controlled. Thus there is really no way of knowing whether the behavioral and physical types concerned (assuming that Sheldon's correlations were accurate) were caused by the same gene complexes; or if social treatment of certain physical types leads to specific behavioral patterns; or if the relation is a simple one between the capacities of certain physical types for certain types or ranges of behavior (tall people make good basketball players, muscular individuals good ditch diggers).

A series of studies over the years beginning in the late nineteenth century have been concerned with

differences in perception among different populations. In many cases, these studies have been carried out on preliterate peoples. Such factors as
response to visual, auditory, and tactile stimuli have
been tested. In general the results have been am-
biguous and sometimes contradictory. In most cases,
genetic and cultural factors were not separated so
that it is impossible to tell how the observed differences were produced. Differences in perceptual
acuity undoubtedly affect behavior so that experiments in this area could, if they were better designed, help to sort out the underlying bases for
certain behavioral differences even if they could not
determine whether the differences were genetic, environmental, or a combination of the two.

In an article in the British journal *Nature*, four
geneticists, Julian Huxley, Ernst Mayr, Humphry
Osmond, and Abram Hoffer, suggested that the form
of mental illness in man known as schizophrenia
has a genetic basis. For obvious reasons, it is impossible to provide experimental evidence for this
(such as cross breeding) but statistics from population studies tend to support the hypothesis. The
authors further suggested that the gene for schizophrenia was dominant but of low penetrance. This
would seem reasonable, since schizophrenia appears
to run in families but does not generally skip generations as would be expected if it was due to a recessive gene. Low penetrance was offered as an
explanation of the fact that the disease occurs in
lower frequencies than would be expected in affected families if the dominant-gene hypothesis
alone was correct. One other fact had to be explained before the suggestion of a genetic basis

could be taken as a reasonable one. Schizophrenia is quite common in most human populations. The high frequency of the suspected gene had to be accounted for. Turning to medical records of known schizophrenics, the authors found that they are highly resistant to certain infectious diseases and to physical stress resulting from severe burns. This resistance might give those individuals carrying the abnormal gene a selective advantage over normal individuals. Actual schizophrenics would have the disadvantage of the infectious disease, but normal carriers, those in whom the gene had no penetrance for the disease, would have all the advantages without the consequences of schizophrenia. Studies of identical or homozygotic twins reared apart tend to confirm the genetic hypothesis, since in many cases each twin of a pair is schizophrenic. Dizygotic, or nonidentical, twins do not show this relationship.

Mitigating against the hypothesis under discussion is the fact that individuals subject to one kind of stress tend to develop resistance to other kinds. Rats rotated in a drum with their limbs tied so that they tend to fall against the surface develop trauma resistance over a period of several days and become less susceptible to electro-shock stress and certain infections. Until recently many schizophrenics were subject to electro-shock therapy which might produce a cross immunity to other types of stress. Statistics on the stress resistance of patients hospitalized since the introduction of chemical therapeutic agents might not show the same correlations as presented by Huxley et al. Thus an environmental factor rather than a genetic one could be responsible for the cited statistics.

At the present time the hypothesis relating schizophrenia to a dominant gene of low penetrance must be taken as suggestive but unproven. Further evidence must be gathered and presented before the scientific community will accept it as a fact of genetics. One would expect, for example, that the frequency of schizophrenia would be higher in populations where physical stress is an important aspect of daily life. In such societies the partially penetrant gene would be more advantageous than in societies where such stress was less common. One would also have to study the social or other environmental conditions which produced high penetrance of the disease on the one hand and low penetrance on the other.

The problems inherent in unraveling the environmental and genetic elements in schizophrenia stand as a model for all studies of human behavioral genetics. It has long been suspected that there are strong hereditary elements in artistic ability and other intellectual skills, but as usual the fact that artists of one type or another tend to have parents who themselves have been successful in artistic pursuits does not help one whit to separate hereditary and environmental factors. Identical twins are rare in the population; identical twins reared apart even rarer. The search for identical twins reared apart both of whom or one of whom is artistic is like searching for the needle in the haystack.

The probability for future successful work in human behavioral genetics has been enhanced recently with the introduction of computer technology. Complex data on behavior and degree of genetic relationship between a wide range of relatives brought up in the same household and apart can be analyzed for degrees of similarity and difference. In this way the limitations of twin studies may be partially overcome.

Genetically Controlled Complex Behavior

The behavioral units I have been discussing thus far have been relatively simple—that is, gene action has been related to a single more or less discrete trait. It should only take a moment of reflection, however, to realize that genetically based behavior can be much more complex. The web spinning of the common garden spider, for example, requires an exceedingly complicated series of moves for the completion of the activity. These moves are built into the genetic code of the spider as much as the number of legs or the color of its body. Different spiders spin characteristically different webs, and within any one species variations in web shape can be related to genetic variation.

The hog-nosed snake, or puff adder, presents an amusing example of complex behaviors related to protection and escape. When disturbed the snake coils itself up and puffs out its neck region like a king cobra. The puff adder may hiss and feint, striking at the intruder, although the snake is not poisonous and rarely bites. If this display fails to impress its audience, the snake will then go into convulsions and finally flop over on its back with its mouth gaping wide. To all appearances it has died of fright. If, however, the snake is turned over on its belly, it will repeat the entire performance.

Complex patterns of behavior are common to

many animals, and in each case they are as species-specific as any series of somatic traits. Nesting behavior in birds varies in terms of materials chosen, architecture, choice of site, nest size, and the time of year in which construction takes place. In some bird species only the males build nests; in others, females; and in still others it is a cooperative activity engaged in by the mating pair. In the hornbill of Southeast Asia, the male encloses a nesting female in a hollow tree trunk, plastering her in with mud. He leaves a small hole through which he feeds her during her confinement, and she does not emerge until the eggs have hatched and the young developed.

Courting behavior, including dramatic sexual displays also common in birds, are again species-specific and are usually innate patterns. Complex mating behavior is widely distributed and is found in such divergent forms as fish, reptiles, and mammals as well as birds. These behavioral systems are as important for the survival of the species as strong beaks in hawks and webbed feet in ducks. Such systems can be grouped as goal-directed activities which include: self-protection, mate protection, protection of the young, courting, nesting, food getting, and in many species social grooming, which has a good deal to do with the social aspect of animal populations (which is itself an area for genetic investigation). The effectiveness of a particular animal in carrying out any of these behaviors can affect its chances of producing offspring. Such patterns therefore have obvious selective value, just as do opposable thumbs or stereoscopic vision in primates.

Environmental Modification of Behavioral Genes

Behavioral genes are subject to all the rules which apply to somatic genes. Mutations occur which alter behavior; some traits are based on dominant genes, others on recessives, still others on codominance. Some traits are single-gene effects; others, particularly complex behavioral systems, are polygenetic. The environmental background on which behavioral genes operate can influence their expression or suppress them.

Within the last few years, environmental manipulation of Tryon's S_1 and S_3 rats has demonstrated the plasticity of genetically determined behavior. If, for example, a group of S_3 rats, inbred to obtain animals as genetically identical as possible, are raised in two kinds of environment, the results show behavioral differences as highly variant as any shown by genes. One group of rats was raised in enclosed cages with no view of, or contact with, other laboratory animals. These rats were never handled by laboratory personnel, and were denied such stimulation as circular runs or objects in the cage which could be manipulated. The other group of rats was raised in an "enriched" environment. They could see other rats in adjoining cages had objects to play with and were able to exercise in runs at will. After maturation the two groups of rats were tested for maze-running ability. The rats raised in the deprived environment performed very badly compared to those which had been exposed to a variety of stimuli during their periods of maturation. Physiological tests run to determine just what physical changes occur when these animals are raised in different environments are now in progress. Those which have been completed suggest that brain development is stimulated or retarded by the degree of stimulation available during the maturation period.

While I hesitate to extrapolate from rats to humans, these results are suggestive of similar findings by contemporary sociologists. Recent studies of children indicate that environmental deprivation may have permanent effects on intellectual capacities. It would appear from these studies that high or low I.Q. is the combined result of heredity and environmental effects. This is in keeping with a major point of this book; that is, that all phenotypic traits are the end product of genetic and environmental interaction. It is logical that the expected relationship between intellectual performance and environment would be strong in humans. The human infant is born very immature, particularly in the structure of the central nervous system. It would seem that the development of the central nervous system is dependent upon genetic potential and a range of external stimuli which produce a complex network of neural connections in the maturing individual. While no one should think of the brain as being anything like a muscle, brains do appear to need considerable exercise for the full development of their genetic potential.

That certain behavioral genes can be suppressed or strongly modified by environmental effects has also been demonstrated in dogs. While cocker spaniels have a genetic tendency for docility, they can be made aggressive through training, and conversely, docile basenjis may not be born often, but they can be made docile. Captive wild animals brought up in the company of man behave very much like domestic species and can become highly affectionate companions. Elsa of *Born Free,* the African lion brought up in the company of humans, is a recent and dramatic example of such behavioral modification of a wild species. Fearing that Elsa would become a victim of ignorant hunters, her human friends had to train her to retake her place among wild lions.

Certain patterns of behavior are more difficult than others to suppress. It is almost impossible to rid cats of their curiosity or minks of their aggressiveness. Sometimes animal trainers find that particular subjects who have progressed well in training revert to more normal patterns which are coded into their genetic structures.

Keller and Marian Breland, professional animal trainers, published an interesting and amusing article on this problem in a 1961 issue of *American Psychologist.* Attempting to train a raccoon to pick up coins and put them in a box, the Brelands found that the animal had difficulty letting go of the coin. After some time the raccoon began to rub the coins together and to dip them into the container. "He carried on this behavior to such an extent that the practical application we had in mind—a display featuring a raccoon putting money in a piggy bank—simply was not feasible. The rubbing behavior became worse and worse as time went on, in spite of nonreinforcement."

In another training attempt, pigs, instead of learning to pick up dollar bills and carry them to a bank, began to drop the bills and root them. In an attempt to correct the behavior, the Brelands increased the animals' drive by depriving them of food and as usual by rewarding only correct responses. Although pigs usually learn easily, in this particular type of training the pigs consistently reverted to a "normal" pig behavior.

The Brelands explain this type of response as "instinctive drift."

"The general principle seems to be that wherever an animal has strong instinctive behaviors in the area of the conditioned response, after continued running the organism will drift toward the instinctive behavior to the detriment of the conditioned behavior and even to the delay or preclusion of the reinforcement. In a very boiled-down, simplified form, it might be stated as learned behavior drifts toward instinctive behavior."

Inborn patterns enable animals to operate effectively in their normal environmental niches. A disturbing influence in the environment can destroy the species if normally appropriate patterns of behavior become inappropriate. Male moths responding positively to light stimulation are burned to death when they come too close to electric lights. Before the advent of artificial illumination, however, light stimulation may have been an adaptive pattern which stimulated activity in male moths so that they might find the more sedentary females. Ducks flying toward decoys are responding in a nonadaptive way to a normal social pattern.

The relative inflexibility of these patterns has recently been exploited by scientists interested in the destruction of certain insect pests. Environmental cues which trigger specific behaviors are turned against the insects. Sound recordings of female mosquitoes, for example, can lure males to their death, or artificial scents which mimic sexually distinctive odors of species can draw prospective mates into waiting traps.

VII
BEHAVIORAL EVOLUTION

Any system which is self-regulating must by definition be one in which the supporting units of the system can change their value. These changes usually occur in response to environmental variation and act to maintain the stability or integrity of the system. Changes which occur in these units can be referred to as behavior. The sum total of integrated behaviors displayed by these units can be referred to as the behavior of the system.

A good deal of energy can be expended by biological systems merely to maintain stability. This kind of activity is analogous to a man on a treadmill who, in order to stay in one place, must move rapidly against the direction of the machine. As we have already seen, temperature regulation in warmblooded animals is a complex process which does nothing more than keep an organism within an acceptable temperature range. To a naïve observer nothing happens to a sleeping dog, yet its metabolic system is busy regulating temperature as well as other vital processes. Other behavior can be more dramatic and open to direct observation. When a hungry dog is presented with adequate food, an observer will have no difficulty in describing a chain of activity which leads to satiation. For the moment we can separate the two kinds of behavior discussed thus far by saying that one kind acts to maintain a permanent steady state, while the other kind acts to change the state of the system in some way. Regular changes in system states can be placed in such ordered sequences as the following: asleep—awake—asleep; hungry—satiated—hungry; sexually excited—sexually satiated; etc. A series of specific behaviors can be associated with such changes in system states, and these behaviors can be linked to variables in the system under investigation. Adaptive behavior carries the organism through a series of activities which fulfill some need. Some psychologists would say that behavior which changes the state of the system in the direction of need satisfaction is initiated by a *drive,* and that the drive itself is initiated by some imbalance in the system. As far as the maintenance of the system is concerned, successful behavior results in need fulfillment, or, in psychological terms, *drive reduction.* Seen this way, there is actually little difference between the two classes of behavior under discussion. Drive reduction means return to some optimal-system state, and drive-initiated behavior is therefore an aspect of self-regulation. Fluctuations in temperature may be less dramatic than some other fluctuations in systems, but the adaptive significance of all behavior is the same, the maintenance of the system.

Prerequisites of Adaptive Behavior

In the last chapter I discussed the genetic basis of certain behavioral sequences. In this chapter I should like to examine certain prerequisites for adaptive behavior as well as the difference between genetically based, or innate, behavior and learning. From now on my attention will focus only on animals, since our major interest is directed toward processes which have contributed to the development of human behavior, and humans share many characteristics with other animal species.

Behavioral evolution has the same restrictions as somatic evolution. All adaptations are environment specific. There are aspects of behavioral development which make adaptation to a wider range of environments possible, just as there are somatic adaptations which widen the range of exploitable niches. The development of lungs, for example, which began in fish, opened the land to vertebrates, and the development of warm blood and homeothermic processes allowed vertebrates to expand into cold climates efficiently. Somatic and behavioral traits which have evolved in relation to a specific environment may maintain their adaptive significance in other environments.

Two major types of adaptive behavior have evolved in the animal kingdom. These are innate responses and learning. With the exception of a very few primitive organisms, both types of behavior play a role in adaptation, but the degree to which an organism's responses are patterned by one or the other varies from species to species.

In simple organisms like the one-celled protozoa, even innate behavior is limited to a small range of responses and a small number of stimuli. In the paramecium, for example, the stimuli to which the animal will react are limited to light, gravity, acid, or gas concentration and, at least under laboratory conditions, electric current. Fine discriminations within each of these categories do not occur, and the response consists simply of an increase of random activity which eventually carries the animal away from some irritating stimulus. This kind of response, known as *kinesis,* can be understood by imagining a simple machine which is capable of movement in any direction and which changes its direction at random. The activity of the machine is dependent only upon the intensity of some stimulus. The machine will speed up and change direction more frequently in regions of high intensity and slow down in regions of low intensity. Under these conditions the machine will eventually come to rest in an area outside the range of stimulation. Such a machine is not very efficient, since it takes a good deal of time for it to get away from some irritant and because this type of behavior does not allow it to react positively to anything. From the point of view of evolution, however, this behavior does offer protection against harmful environmental changes. Since paramecia feed merely by straining bacteria and smaller protozoans out of their environment as they move, positive response mechanisms are not as necessary for survival as those which provide escape from potentially damaging stimuli. Still, such a limited range of behavior would be highly unsatisfactory for organisms which had to actively seek out their food or, in higher forms, their mates.

Even simple multicellular organisms, or metazoa, display a wider range of behaviors than the protozoa. These behaviors are also more efficient than those dependent upon random activity. In Planaria, for example, the increased efficiency in behavior does not display a much higher degree of complexity than that already described. Again, Planaria respond to a limited number of stimuli with a limited range of discrimination. These organisms are capable, however, of responding to positive as well as negative stimuli—that is, of moving directly toward or away from the point of stimulation. Such behavior, known as *taxis*, is the major response pattern of Planaria and other Platyhelminthes (or flatworms), as well as a wide range of invertebrates. A mechanical model of this type of behavior would require a somewhat more complicated machine than is required for responses classified as kinesis. In the case of taxis, discrimination must take place between two classes of stimuli: positive and negative. The machine must now be coded for the appropriate response to each class. Taxis is also a common class of behavior among the Arthropoda, which include the insects, Crustacea, and spiders, but other more complicated behavioral responses also occur in this group.

Anyone who has experienced difficulty in shooing a fly or a bee out of an automobile has observed two kinds of taxis in operation. In both the fly and the bee, light and surfaces act as positive stimuli. The insects cling persistently to the windshield because it is both a source of light and a surface against which the animal will continue to perform exploratory behavior. The response is certainly adaptive under most circumstances. Escape from an enclosed space requires exploration of the boundaries of that space rather than of the interior of the space itself, and light intensity provides a further clue for an effective escape. In the case at hand, if the behavior seems less than adaptive, remember that glass is a recent invention of man, and that under most natural conditions light intensity would provide a useful guide for a trapped insect. If a bee were consistent in behavior of this type, however, it would be unlikely ever to return to its dark hive. In such cases other innate behaviors must take precedence over the photopositive response.

In species which are highly mobile and which are exposed to a wide range of environmental variation, some kind of switching mechanism must evolve so that the appropriate response occurs at the right time. If we return to a mechanical model of behavior, we now require a machine which is not only coded for each class of response but one which must also be coded for a response hierarchy in which the context of the stimulus determines which response is appropriate. Greater adaptability in this case requires greater mechanical complexity as well. In animals this growing complexity of responses is reflected in the evolution of the nervous system, in which switching mechanisms play an important part.

Behavioral Strings

A still higher order of behavior is exhibited by those organisms which are capable of combining automatic behavioral responses into ordered strings or sequences. These sequences result in outcomes which are highly adaptive. A predatory insect may, for example, stalk its prey in a stereotypic fashion, alter its behavior at the appropriate time by pouncing on the victim, alter its behavior again by injecting venom, and alter it once more by eating the prize. The entire sequence may be initiated by a hunger drive, which results from some sensory activity within the animal, but as the behavior unfolds, the insect must respond to a series of appropriate cues.

In many cases such behavioral strings can result in an alteration of the environment to the advantage of the actor. A spider web is an artificial product of behavior which enhances its manufacturer's ability to feed itself. A bird's nest offers "unnatural" protection for its young. In the case of the spider the building material is manufactured in its own body. The behavior is no doubt triggered by some drive, and once begun must proceed according to the blueprint coded into the individual's genetic structure. The amount of external sensory discrimination which is required is probably minimal, since the spider need only connect strands to some supporting material in the environment. Nest building in birds is further complicated by the fact that the bird must choose the appropriate building material. It must function not only with a good internal code system which patterns the construction, but also with preceptors with which it makes correct discriminations.

At this stage our machine is highly complex. Strings of behavior, as well as unit behaviors such as flight from danger, are triggered by internal and external stimuli. Switching mechanisms help determine which response is appropriate to a given situation, but in addition a large vocabulary of responses requires that the machine perceive differences in the environment to a rather fine degree. As the range of behaviors increases, the range of perceived stimuli must also increase.

Learning

A further complication arises when we consider learning. Until now I have been speaking of behaviors which arise automatically in response to certain external and internal stimuli. Organisms which respond in this way are born with a built-in behavioral system. But almost all animal species are capable of learning from their environment. That is, placed in a given context, appropriate responses may develop which are not precoded.

There are many theories of learning and many schools, each with its own point of view as to just how animals learn. Since this book is primarily concerned with evolution, I shall avoid any discussion of the various approaches to this problem. What is of great interest here is the amount of flexibility which learning provides for an organism. The learning process is a source of variation in behavior and is as important to evolution as the development of any somatic trait or any genetically determined behavioral trait.

Within the genetic capability of its species and its own genetic capability, an animal can learn to discriminate between a wide range of environmental variables and to respond appropriately to them. At the same time, an animal can learn to group certain stimuli into classes and thus redefine its environment according to a simplified individually adaptive program. The learning process can completely revise an organism's view of its environment. A pet dog, for example, may learn that people in uniform react differently to him when he is running loose from the way other people react. Or he may come to discriminate between different neighborhood children on the basis of their friendly or cruel behavior. He may know which dogs and cats to avoid and which neighbor will give him a bone. He may learn the boundaries of his master's property and claim it as his own, defending it against other dogs, and unfortunately sometimes the postman as well.

Dogs are quite smart and as domesticated animals they learn a good deal from man about a very special kind of artificial environment. Animals in the wild, however, also benefit from the ability to learn. Young lions practice fighting and hunting with their littermates as well as with their parents. Monkeys learn to avoid certain foods and to eat others. They also learn a wide range of calls which are used by members of their group to signal such things as danger, food, or play time.

Animals can learn directly from their environment, individually experiencing a range of conditions, or they can learn from other animals, particularly if they belong to social groups. Social learning can be patterned so that the behavior of individuals in one monkey troop taken as a whole can differ significantly from the behavior of individuals in another troop.

Learning versus Innate Behavior

One of the great problems in the study of animal behavior is the difficulty in sorting out which behaviors are learned and which are innate. Here, again, there are different schools of thought. Some students of animal behavior believe that there is no such thing as completely innate behavior in higher organisms such as mammals. They believe that there are certain more likely responses given a certain stimulus, and that experience rapidly fixes the appropriate pattern into the nervous system of the animal. Others go far in the other direction and suppose that almost all animal behavior is patterned according to a predetermined code.

The problem is a very difficult one, and again I have no intention to take a stand on it here except to point out certain cautions which have been revealed in the course of recent animal studies.

First of all there appear to be certain behaviors which show an intermediate relationship between inborn traits and learning. Ducks, for example, will *learn* to follow any noisy moving object presented to them shortly after they hatch from the egg. This is called *imprinting*. Under normal circumstances the first object which has the necessary characteristics is their mother, and the response is perfectly adaptive. Observations of ducks in the wild could never reveal the startling fact that the pattern is learned. On the other hand, the timing of the process is under the strict control of the genetic structure of the duck. If young ducks are isolated for a few days and are then presented with an appropriate stimulus (even their own mother) they will not learn to follow no matter how hard the experimenter may try to elicit the expected behavior.

For a long time it was assumed that birdcalls were inborn and that a bird which had been iso-

lated from other members of its species would be
capable of making the appropriate sounds. This is
often true, but it has been demonstrated that in
some species the male will not produce the mating
call unless he is able to learn it from a mature male
bird. In addition to this (as in the case of imprinting) the male must be presented with the learning
model at a certain point in his life or he will never
perform a most important part of the mating ritual.

Comparative Capacities for Learning

The amount an animal can learn, the complexity of
the responses, and the complexity of the stimuli
which can be discriminated also differ according to
species. Certain animals are capable of solving more
difficult problems than others, and the types of
solution attempted by various species also differ.
M. E. Bitterman has studied problem-solving in a
range of animals from fish to rats, and finds that
fish are unable to reverse their ground when a previously rewarded stimulus is no longer rewarded
and a previously unrewarded stimulus is substituted. Rats, on the other hand, soon learn to switch
from one stimulus to the other as the experimenter
varies the conditions. Bitterman also found that
turtles lie somewhere between fish and rats, since
they were able to reverse when presented with
spatial problems (discriminating left from right,
for example) but were unable to reverse when presented with purely visual problems such as the difference between a square and a circle. This occurred in spite of the fact that turtles and fish can
be taught initially to discriminate between a

wide range of spatial cues. In another experiment
Bitterman found that if the reward was varied at
random between two stimuli during the same run
of an experiment, with one stimulus rewarded 70
per cent of the time, rats and pigeons were soon
able to maximize their reward by continually choos-
ing the more frequently rewarded stimulus. Fish
and turtles never adjusted to the experimental conditions and attempted to match by alternating their
responses between the two stimuli. Interestingly
enough, when human subjects are presented with
this kind of problem they often try to "beat the
system"—that is, they try to find out what the actual
pattern of rewards really is. Since the rewarded
stimuli are chosen at random by the experimenter,
there is really no system to discover. A player in
such a type of game will always score higher if he
chooses the more frequently rewarded stimulus on
every turn. Humans evidently are capable of out-
smarting themselves!

Behavior and Morphology

The evolutionary goodness of fit which develops
between a species and its environment includes a
harmony among the traits of the biological system
which has undergone considerable adaptation.
This harmony includes not only a functional architecture, but a harmony between the physical structure and the behavioral traits which develop along
with them in the evolutionary process. Dogs, for
example, are strong-jawed, swift runners capable
of tracking game animals for long distances. Cats
do not have this endurance or long-range speed,

but they are more graceful animals capable of moving silently as they stalk their prey. Dogs are also social animals, hunting in packs, which often bring down a large percentage of grazing animals. Cats usually kill alone and seek out a single animal. The cat is equipped with five weapons: its mouth armed with sharp teeth and four sets of sharp claws. A single cat is a good match for almost any animal up to twice its size. Dogs, on the other hand, have the habit of snapping at their prey and darting away, protecting themselves by feints and dodges. In single combat the dog's prey might well escape, but in a team effort the dog's strategy is quite effective.

Adaptive behavior is a general feature of all animal populations, but the behavioral repertoire of some species is more versatile than that of others. The development and expansion of this repertoire is a genetic process. Behavioral repertoires may be innate or learned or a combination of both. What an animal learns is a function of its genetic potential and its experience. Exactly what is learned depends upon the life situation of each organism, but the type of and capacity for learning is genetically controlled.

Innate behaviors have the advantage of continuity from one generation to another; experience is unnecessary for the development of the appropriate response. A specific environmental cue triggers the behavior automatically and most of the variation that occurs is based on genetic error. Behavioral differences in this case are equivalent to somatic differences within a population and are based on genotypic variation.

Innate responses evolve in species over a long time span and are selected out in the same way as somatic traits. Such behaviors, although adaptive, lack the benefits which may accrue to an organism which can vary its responses in such a way that new or unusual situations can be dealt with successfully. Organisms which can learn are capable of exploiting their environment better than those which cannot. Under stressful conditions they can survive when nonlearners or poor learners could not. Learning is a mechanism of variation which is adaptive for the individual organism within its own life-span. This kind of variation is powerful for the preservation of species, because learning as positive variation is expressed immediately. Natural selection works to preserve good learners as well as new types or capacities for learning as they develop out of the genetic system. The learning process is a maximizing device for organisms because it provides them with a highly effective feedback system in which information from the environment can be evaluated and acted upon in the continuing process of self-regulation.

The Evolution of Behavioral Specializations

All animal behavior requires certain specializations which are related to the development of the nervous system. For it is the nervous system which controls both the reception of incoming signals from the environment and behavioral responses to these signals. This aspect of nervous system activity can be divided into *afferent* responses, or perceptual activity, and *efferent* responses, or reactivity. Afferent responses relay information about the environment to the activity centers of the organism so that it can respond to appropriate cues.

Perceptual discrimination varies tremendously from species to species. This variation includes the type of perceptions emphasized and the quality of the discriminations possible. Mosquitoes and certain fleas are highly sensitive to the carbon dioxide exuded by their prey. Leeches are sensitive to blood in unbelievably minute dilutions. Ticks can be activated by slight changes in temperature which signal their hosts. Birds of prey, such as hawks and eagles, have exceptionally keen vision, and members of the cat and dog families have developed olfactory senses which are powerful, not only in terms of discrimination between scents, but also in the ability to pick up extremely weak scent signals from the environment. One of the major developments which occurred in the primate radiation which led to man was a de-emphasis on smell and a development of stereoscopic color vision congruent with the arboreal life of most primates. Imagine the difficulties which would be encountered by a highly mobile, tree-living animal which could not judge distances as it leapt from branch to branch!

The type of perceptual mechanisms developed in particular species is a clear indication of specialization to particular environments.

Responses to perceptual cues generally require some major adjustive reactions on the part of the whole organism. Mobility is an obvious, but nonetheless major, form of adaptation in the animal kingdom. The kind and the degree of mobility

which species exhibit are just as much forms of
specialization as is the development of perception.
Among birds, for example, gulls and hawks are
able to swoop down directly on their prey from
great heights. They are also capable of tremendous
bursts of speed as well as a characteristic hovering
flight during which they can search the ground for
potential victims. Swallows, on the other hand,
have a characteristic flight pattern which is adaptive for picking flying insects out of the air. Members of the deer and antelope family poorly
equipped for defense are equipped with highly
developed senses of smell and hearing and swift
legs which can carry them away from their
enemies. Members of the cat family have powerful
muscles in their hind legs which enable them to
spring instantly and accurately at their prey. Humans use their hands and minds in combination to
produce an almost unlimited range of behaviors.
Aided by excellent powers of discrimination, a fine
combination of fingers with an opposable thumb,
humans are able to manipulate even the finest
machinery. Man's bipedal gait is certainly partially
responsible for what we call human nature, for it
freed the hands from locomotive activity to become
the servant of the mind.

The type of behavioral activity which is characteristic for an animal depends upon a successful
combination of perceptual discriminations and responses. The perceptual system of the organism
must distinguish between cues, including the
proper context for the cue, and noise in the environment so that the behavior is triggered at the
right moment. In innate systems this is a relatively

automatic process. The presentation of a cue in its proper context sets off what must be something like a chemical reaction in the organism, which in turn produces the behavior. In learned responses the organism may be rewarded when it reacts appropriately to certain cues or punished when it reacts inappropriately. After a certain number of trials, the number depending upon the learning capacity of the organism, the behavior becomes associated with the cue. The behavioral system of the organism becomes tuned to the environment. Since each event which occurs in the environment of the organism is different from every other event, the organism must also learn to generalize in such a way that a series of discrete events become members of one class which can be distinguished from other classes. To do this the organism must distinguish between distinctive attributes of the events or cues and noisy or inappropriate attributes. Thus, if in a behavioral experiment an animal is presented with a square and a circle and rewarded only when it reacts to the square, it must distinguish between two geometric forms—it must learn that the square brings a reward. If the animal is capable of generalizing, it can then be trained to choose any square over any circle. The experimenter could then vary the size or the colors of the cues. If any square is the appropriate cue, then the size and color of the geometric form would be noisy attributes and not relevant for the appropriate behavior. On the other hand, the cue could be made more complicated by requiring the animal to distinguish between a red square and all other forms, including blue and white squares. In this case the color and the geometric form are both distinctive attributes, and the size is a noisy attribute. The ability of animals to make these discriminations depends upon the fineness of their perceptual mechanisms (for example, can they discriminate color at all?) and their innate ability to notice or to sort out the various distinctive attributes from those which are irrelevant to the problem.

Concentration, Memory, and Abstraction

Complex behaviors also require that the organism be able to concentrate during the course of activity. In innate behaviors the environmental cues apparently trigger reactions which can then themselves become cues for further appropriate activity until the entire sequence has been completed. Intellectual activity in higher animals such as primates including man requires that the interest of the animal in what it is doing be maintained and that the number of elements in the behavior be small enough to remain integrated in terms of the over-all objective of the behavior. The animal must also remember the appropriate responses. The number of elements varies from individual to individual according to innate ability (intelligence and capacity) as well as on the previous training. Children are born with immature nervous systems. There are certain tasks which are too complex for them to complete successfully or which require too long a time for their completion. As the individual matures, the complexity of the problem may be increased. Humans develop the ability to handle complex problems as they develop the powers of concentration necessary for successful outcomes. One learns not only the rules of chess, but how to develop successful combinations of plays. This requires thinking ahead to possible outcomes, including what possible changes in the environment might occur, in this case possible moves of the opponent. The player must keep in mind the goal, a checkmate, his strategy, and the possible strategy of his opponent. Humans can learn more than other animals because they are able to make fine discriminations between situations, because they are able to learn new combinations of distinctive attributes rapidly and retain them in memory, and because they are capable of thinking about things in the abstract. The capacity for abstraction is probably distinctly human. Humans can think about things which are not actually happening. The environmental cue system can be constructed artificially within the mind of the thinking individual.

Language

One can think of a beefsteak and how to prepare it even in the absence of hunger, and one can also think out a complex mathematical problem. The major tool for this kind of high-order behavior is a symbol system, or language. Language involves an arbitrary connection between a thinkable symbol, most commonly a combination of sounds, and some object or concept. Thus the sounds (g), (o), and (d) can be combined to form the word or concept "god," and the same sounds can be combined to form another word or concept "dog." The fact that these sound combinations are arbitrary is easily demonstrated by examining the sound combinations which stand for these words in other languages. In French, for example, the sounds (d), (j), and (ø) signal "god" (*dieu*), and the sounds (š), (j), and (ε) signal "dog" (*chien*). Notice that it is not the sounds alone which determine the particular concept, but the order in which these sounds occur. "Dog" and "god" contain the same sounds, but a change in placement changes the concept. This adds a certain economy to language. A limited range of sounds can be arranged into a much larger set of combinations to produce all the symbols necessary for communication and thought.

Of course language consists of more than words. Single concepts can be combined to produce strings of thought or sentences. All languages have sound rules, which determine the construction of words, and rules of syntax, which determine the ways in which words themselves can be combined to produce complex thoughts or complex behavioral strings. The thought "I will wear rubbers today if it rains" (which can be expressed, by the way, whether or not it rains) has a meaning determined by the words and their order within the sentence. The thought expressed by this simple combination of words reflects a potentially adjustive behavior on the part of the thinking individual who will produce a behavior if the environmental circumstances are appropriate for that behavior. "I will do something appropriate [wear rubbers] if something happens [if it rains]." An individual can prepare himself for a possible event by thinking ahead. Language in combination with learning about the virtues of wearing rubbers on rainy days enables the human to respond appropriately to a situation. Thus the capacity for learning language is a highly adaptive trait and one which is distinctively human.

The fact that languages are shared with other individuals, members of the same speech community, means that individuals can communicate not only with themselves as in individual problem-solving but with others and thus engage in cooperative or even competitive activities. Thus one can ask another individual to "Help me lift this log which is too heavy for me to lift alone" or one can say, "If you cross this line I will punch you in the nose." A shared language enables human beings to go through a series of adjustive responses without doing anything more than talk. Thus the sequence which follows from "Let's not fight over this but talk it out instead" has an adaptive value, at least when the talk session leads to some kind of accommodation among the individuals involved.

Language also increases the learning capacity of individuals as well as speeding up the time it takes to learn something. A teacher can present his pupils with arbitrary cases or teach a new technique of behavior rapidly through the communication process. He need not wait until an appropriate situation presents itself to the learner. The situation can be constructed artificially in the teacher-learner interaction. Language enables individuals to learn behavioral patterns of other individuals so that successful predictions can be made about the potential actions of others in particular contexts. Through the code of language humans learn another code, the code of behavior which anthropologists call culture. This code, which is also specifically human, enables men to live and develop as members of highly differentiated cooperative groups.

Preconditions for Human Behavior

In an interesting and concise article, "Somatic paths to culture," J. N. Spuhler of the University of Michigan sorts out the preconditions for human behavior based on the cultural code. All of these preconditions are clearly the outcome of the evolutionary process and demonstrate a complex set of interactions between somatic and behavioral development. Spuhler lists seven preconditions for the beginning of culture. These are:

1. Accommodative vision
2. Bipedal locomotion
3. Manipulation
4. Carnivorous-omnivorous diet
5. Cortical control of sexual behavior
6. Vocal communication
7. Expansion of the association areas in the cerebral cortex

Some of these have already been discussed in this chapter. The importance of diet as explained by Spuhler is related to efficiency of animal food in meeting energy requirements and the cooperative nature of the hunting and food-sharing patterns which must have developed along with meat eating in the kind of primate from which man developed. To quote Spuhler:

"Compact animal protein high in calories is a good basis for food sharing. Of nonhuman mammals it is only the carnivores that share gathered food. It is unlikely that the long dependency of human children—so important to the acquisition of culture by individuals—could develop in a society without food sharing. And the amount of information which needs to be transduced in a communication system for plant eaters like the gibbons is small compared to that needed in group-hunting of large animals. Gibbons share, by vocal communication, knowledge about the location of food collected and eaten individually on the site; hominoids share in the location, collection, and consumption of food."

The cortical control of sexual behavior is important because:

"An important adaptation for culture is the change from built-in nervous pathways to neural connections over association areas (where learning and symboling can be involved) in the physiological control of activities like sleep, play, and sex. Cortical rather than gonadal control of female sexual receptivity may not be essential to the hominoid family (observations on other animals suggest not), but cortical dominance in sexual activity may have contributed to the easy transition of the family from a social unit where sex and reproduction were more important than food economy to a unit where subsistence is the dominant familial function."

It should be noted here that Spuhler may be overemphasizing the role of sex in familial organization in primates other than man and minimizing the sexual role in the human family. George Schaller of the University of Wisconsin has found that gorillas are less than sexy animals, and Freud has based an entire theory of human psychopathology on sexual motivation. Nonetheless, the development of cortical control of sexual behavior provides animals so endowed with a wide range of behavioral response potentials which can be worked into a developing social system largely dependent upon culture.

The expansion of the cerebral cortex and the reordering of cortical structures are the basis of all high-order thinking, including language. Man's brain is the basis of his major adaptation, culture.

Outline of Human Evolution

We are now in a position to sketch the physical and behavioral evolution of *Homo sapiens* from his primate ancestors. The reader who is interested in the details of this process should consult the bibliography under this chapter heading.

There is generally little controversy among paleontologists concerning the broad path of human evolution. There are, however, controversies of a technical sort concerning each fossil stage. Each new fossil discovery forces a readjustment of views and sharpens our understanding of the over-all process. In this outline I shall follow the work of Clifford J. Jolly on early hominid evolution. The analysis of later stages represents my view and synthesis (as a nonpaleontologist) of the work of many scholars who do not always agree.

In recent years the major focus of theories concerning the development of culture and hence man has been on the *Australopithecus* group found in east and south Africa. At least some of these fossils (which date back as far as four million years) are found in association with tools manufactured from both stone and bone. It is also clear from their skeletal remains that at least some of them walked upright, for their pelvic and leg bones of one type are very similar in structure to our own although they were much smaller and lighter creatures than we. At the present time the Australopithecinae may be divided into two groups known, because of their relative size and physical make-up, as *robustus* and *gracilis*. Many authorities now agree that *gracilis*, but not *robustus*, was the tool user and that therefore *gracilis* alone lies on the main line of human evolution. It should be mentioned in passing that one authority, C. L. Brace, believes that the two types merely represent males and females of the same species. Another authority, Richard Leakey,[1] who along with his father, the late Louis S. B. Leakey, has discovered several specimens of both types, now believes that *robustus* was a knuckle walker like the modern great apes.

The discovery of the Australopithecinae group and the confirmation that some of them used manufactured tools led to speculation about their role

[1] At the time of this writing, Richard Leakey dropped a bombshell on the paleontological community. He has released partial data on a skull from Lake Rudolf in Kenya which he claims is at least 2.6 million years old and which is more progressive (i.e. manlike) than either *Australopithecus* or *Homo erectus*. Leakey claims a cranial capacity of 880 cc for the specimen. He puts it with the genus *Australopithecus* and *erectus* from the main line of human evolution. The skull was found in very poor condition and had to be reconstructed from many fragments. Its final place in the hominid sequence awaits further study and debate.

in the development of culture. S. L. Washburn and I. DeVore have published several articles suggesting a feedback relationship between upright posture, cerebral development, and culture. In their work they hypothesize that the first major development along the line of *Homo sapiens* was the development of upright posture and bipedal locomotion, which freed the hand for carrying and work. Under these conditions, tools emerged leading to a reduction in the size of the face and teeth, which in turn, along with selection pressures for higher intelligence, led to an increase in the size of the brain and brain-bearing area of the skull. Once the process got underway, better tools would provide a selective advantage and a feedback relationship would develop between better organized and bigger brains and tool using. R. Holloway, an expert on cerebral development in primates and fossil man, has stressed the fact that early hominid fossils already show significant reorganization in cerebral structure.

The picture of *Australopithecus* (or at least the *gracilis* form) presented here suggests that it was they who developed the habit of meat eating and the consequent traits outlined by Spuhler.

However, the discovery of an earlier fossil has led some authorities to question a part of this theory. The fossil in question is called *Ramapithecus* (*Kenyapithecus* in Africa) and it dates back about ten million years. *Ramapithecus* may well represent one of the earliest fossils which separate the hominid line from the pongid (or ape) line. The most important feature of this fossil group is the structure of the dental arch and the teeth which show a definite trend in the direction of hominization, including a reduction in the canine teeth and a flattening of the molars. Judging from *Ramapithecus*, the evolution of the hominid dental arch and teeth *preceded* culture. Clifford Jolly, looking at both the fossil material and living monkeys, concludes that this change was due to a shift in diet from fruit and grass-stem eating to seed eating. Seed eating requires front teeth capable of separating the seeds from their stems and back teeth capable of grinding. Seeds are high in protein and are, when they can be exploited, an excellent food source. Jolly goes on to suggest that a slight desiccation in the climate of Africa, an event known to have occurred, could have shifted the diet slightly in the direction of mixed seed and meat eating. Thus, while Jolly explains the change in tooth pattern differently from Washburn and DeVore, his theory fits in with part of theirs. If the two theories are put together, a more coherent picture of the first phases of human evolution emerges. The first change came about from a dietary shift in the direction of seed eating and affected only the dental pattern. The second shift came with meat eating and tool using and is associated with upright posture and bipedal locomotion. Tools in place of sharp canines served in several capacities—from hunting weapons to knives used in the stripping of tough hides away from the raw meat which became an increasing part of the diet. It is possible too that a division of labor between the sexes developed during the *Australopithecus* stage or shortly thereafter, since hunting alone is usually not a sufficient subsistence mode for humans. In most hunting and gathering societies known today, women gather and men hunt. In general, women supply surplus calories used by the men in hunting activities that are strenuous and require the expenditure of considerable effort. Game serves as a rich protein supply, but cannot supply all the energy needed by a successful group.

It is generally agreed that the stage following the Australopithecinae was occupied by the *erectus* group. These fossils, which date from about five hundred thousand years ago, represent a considerable advance over the previous stage. Although first found in Java (*Pithecanthropus erectus*) and China (*Sinanthropus pekinensis,*) *erectus* specimens or the tools that mark their culture are now known from Africa and Europe as well. While the Australopithecinae made crude tools from pebbles chipped along one edge, *Homo erectus* made well-shaped "hand axes" by chipping a stone core on both sides to form a sharp point. *Homo erectus* had a brain capacity closer to man than to the great apes and is distinguished by large brow ridges which project over the eye sockets and a low, sloping forehead. It is possible that *erectus* was the first member of the genus *Homo* to use fire.

The next documented stage in hominid evolution is represented by a rich array of fossils dating from about fifty thousand years ago. (Thus there is a gap of about four hundred thousand years.) For some scholars this is the Neanderthal group; for others it consists of pre-Neanderthaloids, progressive Neanderthals, and classic Neanderthals. It is possible that the large quantity of available fossils has fed the controversy. At any rate, the argument centers over the place of what for some appears to be a subpopulation of Neanderthals from western Europe. The individuals of the group are claimed to display morphological features that distinguish them from Neanderthals found elsewhere. Their features include extremely large brow ridges and a long, low skull vault. Several years ago Clark Howell hypothesized that the so-called "classic Neanderthals" represent an isolate population which developed when the group was cut off from the rest of the European gene pool during the last glacial period. Recent work on Neanderthal populations suggests that they were a highly variable group and that the distinction between classic and progressive types may have little or no value. A few paleontologists believe that the entire Neanderthal stage was preceded by a much more modern like group, perhaps even an early form of *Homo sapiens*, which later spread through Europe replacing the Neanderthal population. C. L. Brace has argued eloquently against this position, insisting that Neanderthals are subspecies of *Homo sapiens* and occupy a position on the main line of human evolution just prior to the full emergence of modern *Homo sapiens*. These two forms are now usually distinguished taxonomically as *Homo sapiens neanderthalensis* and *Homo sapiens sapiens*.

The development of the species *Homo sapiens* from prehominid ancestors can be charted as follows:

SUBHUMAN PRIMATE EVOLUTION	PHYSICAL CHANGE	BEHAVIORAL CHANGE
Early stage	Beginning of stereoscopic color vision; orthograde (trunk erect) posture at rest; flexible limbs; development of opposable thumb, experiments with a wide range of locomotion (hopping, leaping, jumping).	Use of vision over smell; manipulation of objects with the hands.
Later stage	Flexible face capable of emotional expression; further experiments in locomotion, including knuckle walking (apes) and brachiation (apes and some monkeys); increases in cranial capacity.	Wide range of social patterns according to ecological specialization; return to the ground of some forms; crude tool use at least by chimps.
HOMINID EVOLUTION *Ramapithecus*	Change in dental structure.	Possible change in diet (seed eating).
Australopithecus (probably only *gracilus*)	Upright posture; bipedal locomotion; changes in cerebral organization and brain size.	Use of tools; adoption of meat eating as a major but not exclusive form of diet; possible beginning of advanced forms of verbal communication, but not language; possible division of labor between sexes with economic co-operation within the group.

SUBHUMAN PRIMATE EVOLUTION	PHYSICAL CHANGE	BEHAVIORAL CHANGE
Australopithecus?	Changes in fertility cycle of female.	Constant receptivity of female as sexual partner.
Homo erectus	Increase in over-all body size; increase and changes in brain organization.	More complex culture; language?
NEANDERTHAL *Homo neander-thalensis*	Increase in brain size and cerebral organization; further changes in skull.	More complex culture.

Some authors, particularly Konrad Lorenz, Robert Ardrey, Lionel Tiger, and Desmond Morris, have attempted in different ways to link innate behavior patterns found in lower animals and particularly in primates with human behavior. In my opinion, such theories should be approached with extreme caution. As we have seen, human evolution consists of the development of the capacity for culture, a form of adaptation that allows for great flexibility of behavior based primarily on learning. Lorenz and Ardrey have suggested that territoriality and aggression are major and widespread genetic adaptations and that they account for human behavioral patterns, particularly war and property relations. While I have answered these arguments at length in my book *The Human Imperative* (1972), let me state here that the case made for instinctive aggression and territoriality in man is a poor one. Territorial and aggressive patterns in our primate relatives vary widely; there is no specific primate configuration. When it comes

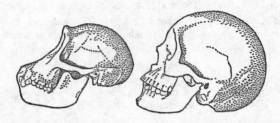

Figure 38. Fossil types.
a. Homo sapiens (*modern man*) *and chimpanzee.*

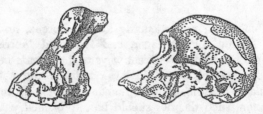

b. Australopithecus robustus *and* Australopithecus gracilis.

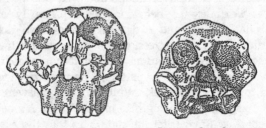

c. Australopithecus robustus *and* Australopithecus gracilis.

d. Homo sapiens neanderthalensis (*classic type from
Le Moustier; part of cheek-
bone missing from cast*).

e. Homo sapiens neanderthalensis (*progressive type,
Skull 5*). *Note the chin,
rounded head, and large
brow ridges.*

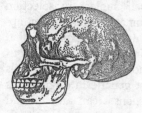

f. Homo erectus (pekinensis). *g.* Homo rhodesiensis.

h. Steinheim man.

to humans, we find that the most technologically primitive groups are the least territorial and many are unaggressive. Furthermore, war is a social phenomenon not a psychological one.

This is not to say that all humans are nonaggressive or nonterritorial. Patterns of aggression appear widely in human societies as do patterns of territoriality (although the latter can be said to be less common). It seems to me that these patterns can be explained in cultural terms so long as we consider both the historical development of human societies and the environmental pressures that bear upon all of them. Selection does not stop because we have culture. Culture as a major aspect of human behavior is itself subject to selection pressures from the environment. Thus, under appropriate conditions, one might well expect to find behavioral patterns emerging that are *homologous* with genetic patterns found among lower animals but that in humans are not genetic in origin. This is our gain. Behavioral responses can turn on and off with great flexibility in a species that depends upon cultural rather than innate patterns.

Summary

If we were to construct a model of behavioral evolution we could make an arbitrary distinction between the development and requirements of innate and learned behavior. Both types of behavior, however, require:

1. Perceptual discrimination mechanisms which are linked to response centers in the nervous system so that cues can be responded to at the proper

time within the life cycle of the individual organism.

2. Mobility enabling the organism to make a whole system response appropriate to the stimulating cues. Responses of an innate nature depend upon a built-in selectivity to environmental cues and a built-in or precoded response potential on the part of the reacting organism. That is, afferent and efferent responses are precoded and linked into behavioral systems. In learned behavior the response potential of the organism must be at least partially uncoded so that appropriate responses can become patterned according to the experience and capabilities of the organism. Animals which learn must also have some form of memory storage so that what is learned can be retained for further learning and behavior.

In innate and learned behavior the organism must be able to separate crucial cues from other stimuli in the environment. For animals which learn, the ability to make new discriminations and new perceptual organizations becomes necessary.

The development of a symbol system in which arbitrary symbols can stand for objects or concepts greatly enhances the learning process and provides the organism with a capacity for learning a higher-order behavioral code or culture.

VIII
CULTURE AND HUMAN BEHAVIOR

I remember watching a blind student several years ago walking across the campus of a large state university. He guided himself with a cane, tapping it against the sidewalk which ran in spokes from building to building. Although he knew the campus well, on that particular occasion he became distracted for a moment and wandered onto the grass, where he immediately lost all sense of direction. His movements became disorganized as he searched hopelessly for a bit of cement. He became visibly panicked until a passing student came up and led him back to the appropriate path. Once again he was able to continue toward his class unaided.

I was struck by the similarity of this situation to the situation of all human beings who have grown up within a particular social milieu. Out of an incredibly large number of possible ways of living successfully, all normal human beings operate within a narrow framework of convention. This convention is sometimes limiting and perhaps to certain individuals unsatisfying, but it provides a set of rules which act as guidelines for action. One learns not only how to behave in given situations, but also what to expect from others. In addition, one learns how to act in relation to the physical environment. Man's existence as a social animal is conditioned by these rules, and no human group could operate successfully without them. These conventions, which are in many ways so similar to the sidewalk for the blind student, make up what the anthropologist calls culture. Culture is essentially a set of rules for behaving in a human way. Different human groups have different conventions, and so there are cultural differences between peoples, but all members of the human species are part of some social group which can be characterized by its own set of rules.

Human beings who for some reason find themselves in an alien society founder very much like the blind student on the grass. Lacking guidelines for action, they are unable to behave in appropriate ways and are unable to predict the behavior of others. Anthropologists who spend a good part of their lives among peoples with a culture different from their own find that each social setting demands a new orientation. Once this is accomplished, the anthropologist can begin to thread his way with some success through the maze of social relationships which characterize any particular society. Anthropological field work is in many ways like a game in which the anthropologist tests and retests his assumptions about the rules until he becomes at least a moderately good player. This is done in several ways. He may, for example, sit passively with an informant and ask questions. This is not as easy as it might sound because the questions themselves must be appropriate to an unfamiliar situation. The anthropologist must learn what questions to ask and how to ask them. He must also learn whom to ask, since in every society different people are expert in different things. The total system which operates in any society is made up of the contributions of various individuals who perform a wide range of tasks.

Assuming that the field worker has done all this successfully, we find his analysis may still be neither complete nor accurate. The actual behavior of individuals in a society may depart widely from their own conceptions of it. The anthropologist must observe actual behavior and test his assumptions about it whenever possible by interacting with members of the society he is studying. He must also learn to distinguish regularized behavior from idiosyncratic behavior. That is, he must separate the rules of the system from the "noise" which occurs due to inaccurate copying or inappropriate responses. When all of this is accomplished, the anthropologist is then able to state the rules of the system which he has been studying.

The phenomena of culture can be divided into three categories: *material culture, social culture,* and *mental culture.* This tripartite division is not always explicit and I borrow it from the work of Cornelius Osgood, whose explication of these three modes of culture appears in his monographs on the Ingalik Indians of Northern Canada.

Material culture consists of the products of man's creative abilities—his art and his technology. All things made by man, from tools to the products of these tools, fit into this category. If we think of culture in terms of behavioral models, however, we must consider these objects as the products of culture. Material objects are produced according to ideas held by the people who produce them. The model consists of at least two parts: the plan of production or the techniques used in producing these objects from some known raw material, and the idea of what the object will look like when it is finished. If the object is to be used, the use is also a part of the idea system associated with the object. Nothing produced by a craftsman will be exactly like any other object produced by the same or other craftsmen. The differences between objects may be due to a limited amount of variation or creativity allowed to the individual, or it may be due to technical factors involved in the production. Anthropologists interested in material culture attempt to discover what rules exist for their production and what the limits of variation are.

The next segment of culture (social culture) involves patterns of action. The anthropologist can observe these in the daily intercourse between individuals in a society and can attempt to derive behavioral rules from these observations. Social culture includes such activities as greeting patterns, systems of rights and duties which individuals exhibit in their relationships with others, the behavioral patterns which occur at particular ceremonies, eating arrangements, the events which occur in legal disputes, etc. Social behavior in human groups is always patterned in terms of specific defined relationships between individuals, between individuals and groups, and between groups. Individuals may be said to occupy a series of *status* positions in society which are coterminous with "correct" behaviors. These behaviors involve rights which accrue to the status as well as duties which such individuals must perform in relation to others. A chief, for example, has certain rights of appropriation and deference in relation to his subjects, but he is also obligated to serve them in some way. In many societies chiefs are presented with gifts of food which they are then obligated to return to the people in certain prescribed amounts.

An individual always occupies several statuses in his society. A man may be a son, a father, a husband, a hunter, and a medicine man. Within the context of the family he may perform duties related to two or more of these statuses at the same time or he may perform the duties of a single status depending on the particular social context. All status positions also carry appropriate affects. A father may be stern in one culture, loving in another, or a combination of both in still another. When an individual is involved in the behavior sequence appropriate to a particular status he can be said to be acting out the *role* congruent with that status.

Ceremonial activities can involve a series of appropriate actions of a large number of individuals occupying similar and dissimilar statuses. An anthropologist interested in describing a particular series of social patterns must analyze out the existing status system and the appropriate role behaviors associated with them. Again, there may be variation in role playing from individual to individual. It is one task of the anthropologist to sort out the limits of variation allowed in the particular society under study.

The third area of culture (mental culture) is the hardest to extract from behavioral data. Mental culture includes the belief systems of individuals in a society and actually makes up the corpus of rules which determine the more directly observable material and social culture. No anthropologist will ever get inside the heads of his informants, so statements about mental culture can only be approximations based on the ability of the informants to communicate with the anthropologist and the ability of the anthropologist to extrapolate information from observable phenomena.

Recently there has been a good deal of argument in anthropology over the ability of the researcher to even approximate the code which is employed by alien people. It is well known that the anthropologist constructs his own model of the society under study which enables him to predict, hopefully with success, the behavior of individuals in that culture. But there does not appear to be any good way of determining if the anthropologist's model is equivalent to the actual culture. (There may be more than one solution to the same problem.) This may not, however, be an appropriate question for anthropology, since the object of study is not individual psychology but the operation of social systems.

Certainly from the point of view of evolutionary studies, the actual model or *cognitive map* which determines behavior is irrelevant. The evolutionist is concerned with the interrelationships between actual behavior and environmental parameters. If he has a good working model of the behavioral system, this is enough for research purposes.

Adaptive Rules and Culture

The rule systems with which humans operate are generally adaptive. They make it possible for a group of individuals to survive more successfully than they could either alone or as members of an unorganized horde. The existence of a social tradition enables them to accumulate knowledge about the environment and to pass it down to the next generation so that adaptive behaviors can accumulate through time. The expectations about the behavior of others in given contexts enables them to act cooperatively with a minimum of difficulty, although this does not mean that man's social life exists without stress. There are times when individual desires and cooperative efforts are at odds, and there is no doubt that the existence of social systems places a strain on the individuals within them. The gain, however, generally far outweighs the loss, particularly if we consider the system as a whole rather than some of the individuals within it who may have difficulties adapting.

Included in the rule systems of all cultures are rules that explain how the natural world and even the cultural world operates. These are often ordered into coherent sets or theories. Theories are useful because they enable people to make successful predictions about the behavior or operation of certain phenomena. They also allow for successful change since, by setting a direction, they reduce search behavior from randomness to order. The discovery of penicillin, for example, was accidental, but a theory that accounted for its operation led rapidly to the discovery of many more antibiotics. The germ theory in general led to the rapid development of effective therapeutic agents for a wide range of infectious diseases.

While good theories can speed change, poor theories can inhibit it. Perhaps because theories are systems, people tend to cling to them. The abandonment of a theory leaves a wide hole in a cultural system. Anthropologists have noted for some time that not all behaviors are explained on the basis of a coherent theory. Often when informants are questioned about the reasons why they do something in a certain way, they reply, "We've always done it that way" or "That's the way our grandfathers did it." Such a response can be very frustrating to an anthropologist who is searching for the root causes of behavior. But it might also be informative, for it signals a behavior that is unattached to theory. Such behaviors may be fixed directly as units into a cultural system by rewards and punishments coming from the environment. Thus the "grandfather" response may signal adaptive traits that have developed unconsciously in a culture.

Traits that are unattached to a theory may be more subject to change or abandonment than those that are part of a theory. On the other hand, the abandonment of a theory may lead to more widespread change. This, of course, is an hypothesis about the operation of cultural systems and needs to be tested. Before leaving it, however, it might be worth mentioning two other possible explanations for the "grandfather" response.

People often unconsciously do things that are, from the psychological point of view, anxiety reducing. A nervous habit like touching the hat before pitching a baseball is one example. When told that he always goes through such a motion and asked why, a pitcher might well respond that he has always done it that way, that it insures a good pitch. Such "superstitious," objectively irrelevant actions can become part of behavior through a conditioning process in which a neutral action accidentally becomes part of an effective, rewarded response. More serious are behaviors that unconsciously substitute for some other action. Freud notes in his collected papers that a woman who had a severe tic developed it during the illness of her daughter. In analysis he discovered that during the illness the doctor had told her to remain still at her child's bedside. According to Freud, this woman had ambivalent feelings toward her sick daughter. She loved her but at the same time felt aggressive because of the extreme fatigue she was forced to suffer. Unconsciously she wanted to kill the child, but of course her love overcame this impulse. The tic became a substitute behavior for a more dangerous movement. If this analysis is correct, and it too must remain in the realm of an hypothesis, we might extend it to certain cultural behaviors, particularly rituals in which symbolic violent action might substitute for actual aggression. The reasons behind such rituals might be suppressed and the "grandfather" response might be called into play to explain them.

Finally, there are cases in which behaviors are strictly imbedded in specific sequences or structures. Because they constitute a "grammatical" element in such structures, informants might not be able to explain them out of their specific context. That is, if they don't know the rules of the grammar (a common phenomenon), they won't know how to analyze the activity and might therefore give the "grandfather" response.

There is no reason to assume that only one of the three explanations offered here for the "grandfather" response is correct. All three could operate in different situations. It is also possible that other viable explanations could be devised. In each case, including the ones presented here, testing is necessary to establish their validity.

Biology and Culture

Culture is learned and not inherited. Physiologically, humans are born with the capacity to learn any language and any culture. There is no evidence that any aspect of culture is inherited, although one must not overlook the fact that certain behavioral systems are more compatible with the total physiological and psychological structure of the human organism. It is probably for this reason, among others, that similarities between rather complex systems of behavior appear in widely scattered areas of the world where there is little possibility that these similarities arose as a result of borrowing. The cultural traits which develop develop against the background of biologically determined human nature.

There is also a possibility that there are average physiological differences between human groups which have some effect on a particular line of cultural adaptation. That is, a genetic substratum may affect the probabilities that certain types of behavior will develop. If this is true, and as yet there is no evidence for or against it, one would *not* expect physiological variables to absolutely determine the kind and direction of cultural development. Behaviorally, the human being is highly flexible, and the penetrance of any "behavioral" gene would be subject to strong environmental and cultural modification. If the dog is flexible in this respect, the human organism is much more flexible. In addition, there has been so much interbreeding between human populations that behavioral regularities of the kind found in purebred lines would be unlikely, even impossible. What may emerge with new lines of research is a broad kind of behavioral trend within populations with a great deal of overlapping between populations, so that it would be impossible to characterize any group in terms of traits peculiar to them alone. Differences between the mean averages of various groups would have to be demonstrated, and these would then have to be correlated with cultural differences. Even if this is accomplished, it will be difficult to sort out the direction of causation. There is a great deal of evidence in both anthropology and psychology that cultural factors influence physiology. Finally, let me say in this respect that significant physiological differences between human groups which might affect the nonbiological behavioral system would be much finer in nature than supposed differences in intelligence, for example. Intelligence tests, if they measure anything, measure a gross complex of neurophysiological activity. It is likely that any differences which may exist between human groups of the type under discussion would be finer than the tools available at the present time to measure them. This does not mean that such a search should be abandoned. Vulgar racism can be avoided at the same time as an honest attempt is made to sort out somatically and nonsomatically inherited behavioral traits and the relationships between them.

Culture and Environment

Human behavior is in large part adaptive, although this does not mean that everything humans do is adaptive. This would be a peculiar expectation to say the least, since we know that not all somatic traits or genetically determined behavioral traits are selectively advantageous. Since the code system or culture is a major human adaptation, one should naturally look at various behavioral systems in terms of their selective advantage in specific environments. Traditionally, anthropologists have examined the relationships between subsistence techniques, the size of population they are able to support, and the complexity of the technology associated with them. It is well known, for example, that hunting and gathering societies, except under the most unusual circumstances, are small groups with simple technologies. Agricultural or herding economies are able to support larger groups of people and generally display a more complex technology and social organization than that of the hunters and gatherers. The development of large agricultural and other food surpluses and a rising population enable full-time nonfood-producing specialists, such as smiths and merchants, to develop with a concomitant increase in both technological mastery and a highly differentiated society based in part on economic pursuits. It has been assumed for a long time by many sociologists and anthropologists that the size of the social group has its own effects upon the kind of social organization which develops. Thus it is argued that subsistence techniques have an indirect but powerful effect upon the development of social systems.

These kinds of regularities, however, are of the grossest sort. Less attention has been paid to the actual environmental niche in which behavioral systems develop than one might expect at first reading. This has occurred for two reasons. In the first place, most anthropologists have been concerned more with the effects of culture on environment, as in technological development, than on the relationships which might exist between the environment and man at any given stage of development of culture. This concern is based on the assumption that man makes his own niche by means of culture. Human behavior is seen as an environment-changing device. This is certainly true to a great extent, but what is more important is the fact that no matter how advanced the society, there are still factors in the niche to which a society must accommodate. It is therefore safer to see culture as a form of adaptation rather than as a niche itself. The second reason little fine-grained research has been done lies buried in the history of anthropological theory, and a brief digression into the subject will be necessary to clarify this issue.

Social Evolution

The first major statement on social evolution came in the middle of the nineteenth century from the pen of Herbert Spencer. Spencer was a self-educated philosopher with interests which ranged from physics to sociology. Spencer's access to empirical data was limited, but there is no doubt that he was an original thinker. Even today many of his ideas are worthy of attention. Spencer's theories have a limited empirical basis; nonetheless, it is to him that we are indebted for the first truly general theory of evolution.

Spencer's theory was what has come to be known as unilineal evolution. Spencer attempted to construct an axiomatic system based upon three principles: the indestructibility of matter, the persistence of force, and the continuity of motion. From these three principles Spencer claimed to deduce universal teleological progress in nature, involving development from the simple to the complex, from the unordered to the highly organized. This he called evolution. Since evolution is a law of nature, it must apply universally. Thus it includes the inorganic, the organic, and the "superorganic," i.e. society. I will attempt to show later that the concept of a "superorganic" level of organization creates difficulties for the biological model presented in this book. Spencer's inclusion of human society in an over-all theory of evolution, however, was a major step in the foundation of scientific anthropology.

On the other hand the universality of Spencer's theory has other implications. The expression "survival of the fittest" (said to have been coined by Spencer) does not imply that the environment is a controlling mechanism in evolution. For Spencer progress does not depend upon conditions. Struggle comes after the fact and is useful in the elimination of less adaptive forms, but better forms arise as a matter of course. They are not better in terms of a particular environment but are superior in some sort of absolute sense.

According to Spencer, as society evolves it gains greater and greater control over the inorganic and organic levels of the environment and is thus better able to continue the process of evolution toward higher and higher forms at an increasing rate.

In a later work, *Sociology*, Spencer paid more attention to the scanty empirical evidence available to him at the time and discussed what he considered to be causative agents which contributed to differential social development. Thus he suggests: "Societies grow; while they increase in size they increase in structure. As the structures differentiate, their functions do also. The division of labor makes the social organism like the individual organism."

What is important about Spencer's attempt to analyze social systems is the attention he paid to relationships between various aspects of social systems and culture. He, along with Marx, was one of the first social theorists to talk about correlations between various aspects of human behavior such as economics and social organization.

The first man to define the discipline of anthropology in modern terms was Edward B. Tylor. His books, *Primitive Society* (1871) and *Anthropology* (1896), set the tone for future anthropological research and helped to define the concept of culture as a specifically human characteristic.

Tylor also accepts evolution as a given. For Tylor, evolution occurs in stages which can be reconstructed through the examination of archaeological evidence and by field work among primitive people who stand as cases of arrested development. Living primitives are taken as an accurate mirror of the past condition of more advanced peoples. Evolution is a process which has occurred in society and which proceeds from savagery (hunting and gathering) to civilization (modern nation states). Tylor's criterion for the measurement of advancement is not technology, but rather the ticklish concept of the "good society."

"Civilization may be looked upon as the general improvement of mankind by higher organization of the individual and of society to the end of promoting at once man's goodness, power and happiness. This theoretical civilization does, in no small measure, correspond with actual civilization, as traced by comparing savagery with barbarism and barbarism with modern educated life." Tylor was not immune to the smug complacency which marked the Victorian period.

Tylor did not subscribe to the "superorganic" of Spencer. For him, "Collective social action is the mere resultant of many individual actions." This is an unfortunate simplification of human behavior and is characteristic of Tylor's approach to evolution in which particular culture traits are arranged in sequences totally out of context. Religion, for example, goes through a series of stages in a vacuum. This is in spite of the fact that Tylor was one of the first anthropologists to use statistical techniques in correlating various aspects of social structure with each other. Such studies were purely synchronic. Tylor did not see evolution as a process in which a series of behavioral variables were linked together in some sort of functional unity. He did not look for independent and dependent variables in systems capable of change. In addition, like his contemporaries, he ignored the role of environment as a selective agent.

In spite of his adherence to unilineal theory, Tylor recognized that at least some change comes about through the borrowing of culture traits by one group from another. It was Tylor who developed the concept of *diffusion* and stated certain rules by which anthropologists could determine whether a trait was invented independently or borrowed. He suggested that when two culture traits were similar and sufficiently complex in conception and/or design, the case for independent invention was weakened. The more complex the similarities, the higher the likelihood of diffusion.

The first great American evolutionist-anthropologist was Lewis Henry Morgan. A lawyer and sometime Republican member of the New York state legislature, Morgan practiced anthropology as a hobby. His initial interest in the field developed out of an attempt to write a constitution for a fraternal organization based upon the Iroquois Confederation. Morgan spent a good deal of time on the local reservation and finally completed a study of the tribe. His first book, *The League of the Ho-Do-So-Wa-Ne* (1851), stands as a major contribution to American Indian studies. One aspect of Iroquois society which caught Morgan's attention was their kinship system. Morgan discovered that they used a system of terminology which differed considerably from our own. The term for one's own biological mother was, for example, applied also to mother's sister. Father and father's brother were included in one category. The offspring of father's sister and mother's brother were called by a term which could be equated with our "cousin," but the offspring of mother's sister and father's brother were called by the same terms as ego's own brother and sister. A few years after this discovery, Morgan took a trip into the Western United States, where he collected kinship information from tribes unrelated to the Iroquois. In one of these tribes Morgan found that the pattern of designating kin was identical to that of the Iroquois, although the actual words used were different. This discovery led Morgan to wonder about the types of kinship systems which might exist. Morgan then constructed a kinship questionnaire and sent copies of it to government officials and missionaries. When the data were returned to him, Morgan found that the types of kinship systems distributed around the world could be organized into a rather small number of types. It became apparent to him that some organizing principle must operate in human societies to delimit the range of possible terminological systems. Since kinship obviously has some relationship to family organization and marriage, Morgan suggested that terminology was the direct outcome of family types and marriage rules. As he became more deeply involved in the problem, he came to believe that marriage systems and family organization reflected stages of economic development. Using stages derived from the new field of archaeology, Morgan constructed a series of correlations between economic organization, social structure, and kinship terminologies. Since there was a good deal of evidence to show that early man was a hunter and gatherer and that agriculture and the domestication of animals occurred at a later stage of human development, Morgan set out to fit his correlations into an evolutionary framework. Morgan operated with the same assumptions as Tylor and other early anthropologists. These were that living primitives are present-day representatives of our own archaeological past and that the evolution of human society follows a single path of development. Extrapolating from living cultures, Morgan reconstructed the social organization and kinship systems of peoples who left little more of their culture behind than some stone tools and a few pots.

Following a tripartite scheme, Morgan divided human development into three familiar stages: savagery, barbarism, and civilization; and divided each of these into three substages: lower, middle, and upper. The lower status of savagery ranged from the infancy of the human race to the commencement of the next stage. Middle savagery was marked by the acquisition of a fish subsistence and a knowledge of the use of fire. Upper savagery was marked by the invention of the bow and arrow. Lower barbarism begins with the invention of pottery, and middle barbarism is characterized by the domestication of animals in the Eastern Hemisphere and the cultivation of maize with irrigation in the Western Hemisphere. Upper barbarism is correlated with the smelting of iron and the use of iron tools. Civilization comes upon the scene with the invention of the phonetic alphabet. According to Morgan:

"Each of these periods has a distinct culture and exhibits a mode of life more or less special and peculiar to itself. This specialization of ethnical periods renders it possible to treat a particular society according to its condition of relative advancement, and to make it a subject of independent study and discussion. It does not affect the main result that different tribes and nations, on the same continent, and even of the same linguistic family, are in different conditions at the same time, since for our purpose the condition of each is the material fact, the time being immaterial."

It is clear from the above quote from *Ancient Society* that Morgan was as uninterested in actual historical processes as he was in diffusion. These stages are categories into which any living culture can be placed. They reflect absolute stages of development so that information on one aspect of culture allows the investigator to assume the presence of other correlated aspects. Morgan describes the correlation between these technological periods and family structure as follows:

"We have been accustomed to regard the monogamian family as the form which has always existed; but interrupted in exceptional areas by the patriarchal. . . . Instead of this, the idea of the family has been a growth through successive stages of development, the monogamian being the last in its series of forms. It will be my object to show that it was preceded by more ancient forms which prevailed universally throughout the period of savagery, through the older and into the Middle Period of barbarism; and that neither the monogamian nor the patriarchal can be traced back of the Later Period of barbarism. They were essentially modern. Moreover, they were impossible in ancient society, until an anterior experience under earlier forms in every race of mankind had prepared the way for their introduction.

"Five different and successive forms may now be distinguished, each having an institution of marriage peculiar to itself. They are the following:

"I. The Consanguine Family
It was founded upon the intermarriage of brothers and sisters, own and collateral, in a group.

"II. The Punaluan Family
It was founded upon the intermarriage of several sisters, own and collateral, with each others' husbands, in a group; the joint husbands not being necessarily kinsmen of each other. Also, on the intermarriage of several brothers, own and collateral, with each others' wives, in a group; these wives not being necessarily of kin to each other, although often the case in both instances. In each case the group of men were conjointly married to the group of women.

"III. The Syndyasmian or Pairing Family
It was founded upon marriage between single pairs, but without an exclusive cohabitation. The marriage continued during the pleasure of the parties.

"IV. The Patriarchal Family
It was founded upon the marriage of one man with several wives; followed, in general, by the seclusion of the wives.

"V. The Monogamian Family
It was founded upon marriage between single pairs, with an exclusive cohabitation."

There is no evidence at all that the first two types of marriage ever existed in any culture. The other types, as well as some complications which can be added to them, may be only partially correlated with the economic stage of development of the group in question. Morgan's error in constructing these marriage types was to assume their existence from kinship data alone. Thus, for example, because the Hawaiians call all kinsmen of their parents' generation by terms equivalent to "father" and "mother" and all kinsmen of their own generation by terms applied to their own brothers and sisters, Morgan assumed that this was a clear indication of group marriage. In such a situation he reasoned no one would be able to distinguish a real father or mother and all children of a group marriage would be brothers and sisters within the group. The danger in Morgan's thinking was his willingness to leave empirical data behind and to deduct marriage rules simply from the terms applied to kinsmen.

Nevertheless Morgan's concept of evolution is more useful than Tylor's. Morgan was clearly correct when he saw evolution as a process of interaction in which aspects of social organization were tied to economic pursuits and technological development. He postulated, as did Spencer, some kind of functional unity in evolution, a unity based on the concept of society as a dynamic system.

Antievolutionism

Early in the twentieth century the work of these evolutionists came under severe attack from two sources. In England and among museum-oriented anthropologists in the United States, the idea of diffusion became an obsession. Tylor's influence and the attention which he focused on material culture led English anthropologists to concentrate on the movement and transmission of material traits. Museum anthropologists were most comfortable when dealing with solid objects, and the discovery of even remote similarities in culture traits was taken as evidence of contact. The fact that some cultures were less advanced technologically was accepted as evidence that they lay away from the main lines of diffusion, which for the extreme thinkers in this school emanated from one area only. Egypt was believed to be the cradle of civilization, and lines were traced outward from the Middle East to the remote areas of the world, showing the distribution and diminution of items of high culture as they flowed away from the center of origin. When aspects of social or mental culture were examined, they were treated as if they were tools or pots, static items which could be ripped out of context and catalogued into a neat museum classification.

Another line of attack on the unilineal evolutionists came from Franz Boas and his students. The overtheorizing of the evolutionists disturbed Boas as much as the idea that primary data on the American Indian was disappearing at an alarming rate as the process of culture change swept the country. While Boas did not set himself against theory, he demanded that theoretical formulations be backed up by solid data, and he directed his students to collect that data. Unfortunately, he erred by assuming that the mere collection of data on the culture of exotic peoples would provide the material out of which useful theories could be constructed. Data exists in overwhelming quantities. It is necessary to have some theory before collecting begins so that the scientist can focus on what is important to a particular set of questions. A good deal of the data collected by Boasian anthropologists is frustrating to the modern worker who has specific questions to ask. In all the great wealth of salvaged material there may be no answers. Good science requires a feedback between theory and hard facts. It is possible to err either on the side of theory or on the side of data collection. Filling the mind with unstructured material can be just as debilitating to clear thought as armchair theorizing about the basis of real behavior.

Historically there is probably another reason why cultural evolution, and Morgan in particular, fell into disrepute in the United States. It was Morgan's lot that Friedrich Engels took up his work and employed it as an anthropological justification of Marxian theory. The evolution of family structure which Morgan constructed provided Marx and Engels with data from non-Western society which served as a buttress to theories which they had originally applied to Western civilization. The theory became and continues to serve as the basis of Soviet anthropology. Engels' book *The Origin of the Family, Private Property and the State* stands as the source book for all anthropological theory in the Soviet Union. A strong anti-Marxian and strangely anti-materialist bias in the American university contributed to making Morgan's work suspect because it fit so well into the evolutionary framework of Marxian economics.

Modern Unilinear Evolutionists

There have been two modern protagonists of unilinear evolution, the English archaeologist V. G. Childe and the American ethnologist Leslie White.

Childe saw human development in terms of great technological revolutions which shaped the destiny of mankind. The first revolution came with the inception of culture. The invention of tools began man's successful struggle with nature. One of Childe's early books is titled *Man Makes Himself*. This ability of man to control the environment and to increase its yield is the major theme of Childe's work. The second revolution came with the domestication of plants and animals in which a steady food supply could be more easily assured. Increases in agricultural and animal productivity allowed some men to specialize in nonsubsistence economic pursuits. Settlements were able to support higher populations, and trade between population centers became an important occupation. This led to the urban revolution. The continuing development of culture led finally to the industrial revolution, which stands as the latest-step stage in the reorganization of man's relation to environment and technology. Childe's theories are more satisfactory than those of the early social evolutionists because he concerned himself with the hard data of archaeology. Even so, many of his generalizations not mentioned here do not stand up well under the light of recent evidence.

From a theoretical point of view, Childe is more satisfactory than early evolutionists for several reasons. First of all, he considered the role of environment in cultural development. In his scheme it was no accident that civilization developed in the Middle East. He felt that after the domestication of plants and animals, certain conditions such as limited arable land hemmed in by deserts forced people to live in small, habitable pockets. Productivity and forced increases in population density, coupled with trade between centers which had command over different natural and man-created resources, led to urbanization, the development of a writing system, and the rise of civilization. His second contribution was to wed the concept of diffusion to evolution. Once civilization had been invented in the ecological zone most suited to its genesis, the culture of cities could spread outward to other groups through trade, migrations, and conquest. Thus there would be centers of origin for culture traits related to environmental adaptation, but there was also diffusion of these traits beyond their original boundaries into other areas where they could also flourish.

Third, Childe recognized that evolution had to be measured in terms of one particular quantifiable variable. Technological change and economic "progress" were factors in evolution, but they could not serve as an effective measure. Childe chose increased population size as an indication of evolutionary success. His application of this measure, which is the accepted standard of evolutionary biologists, was applied only generally to the population explosions released by his "revolutions." They were never used in careful studies of populations in particular ecological zones, nor did he more than abstractly compare the population rates of competing populations. This allowed Childe to make low-level generalizations about the development of human society, but he did not extend his method to a careful comparative study of populations in competition. One reason for this is, of course, the fact that Childe was an archaeologist. His influence on ethnologists was not strong, unfortunately, perhaps partly because of the direction which British anthropology was taking while Childe made his major contributions to the field.

English Functionalists

Briefly, English anthropologists had turned under the influence of B. Malinowski and Radcliffe-Brown to the study of social relations. Anthropology became a comparative sociology focused on the interrelationships between various aspects of social behavior as they formed an integrated whole or so-called "functional" system. Malinowski stressed the biological "need fulfilling" function of these systems in his theoretical writing, but his field work and the field work of his followers was not amenable to an analysis of the relationship between biology and culture since the descriptive material offered was almost exclusively within the realm of social and mental culture. Radcliffe-Brown recognized the adaptive nature of human behavior, but was more interested in correlations between various aspects of social structure. His work was similar in many ways to the early work of Morgan in which an attempt was made to relate kinship organization to social rules. Radcliffe-Brown, however, rejected historical and evolutionary explanations and limited his analysis to individual systems existing at the time of analysis. He considered the ethnohistorical writings and evolutionary theorizing of American anthropologists to be mere "conjectural history." The narrowing of interests of the British continued to the point where they began to define themselves as comparative sociologists. In comparing themselves to American anthropologists, they often point out that their major interest is in social networks while the Americans favor culture as their major concept. Analysis of material culture dropped almost completely out of English contributions to ethnological writing. At the present time, for example, it is almost impossible to get even a preliminary view of African material culture from English ethnologists, even though British Africa has been well covered ethnographically.

The emphasis on culture as a central concept in Tylor and the later Americans and its neglect by the British had much the same effect on evolutionary studies. In the former case, as I have already mentioned, it tended to remove environmental variables from the discourse on human behavior, since study was restricted to the development of culture in its own terms. In the latter case, focus on social relations excluded technology and subsistence activities from the realm of discourse except as they related to social structure itself. Few attempts were made to relate them to environmental controls. Interest in such topics as population density, disease, caloric intake, etc., was almost totally neglected, although there were some studies, particularly in Africa, of the relationship of social organization to food production.

Leslie White

The early evolutionists from Spencer to Morgan neglected environment primarily because one could not be a unilineal evolutionist and at the same time focus on environmentally limited differential development. Culture was something which evolved on its own terms, according to its own rules.

This latter view, rejected only by Childe, has been supported by Leslie White in America. White is willing to admit that local conditions do affect particular lines of development for limited time periods, but his major interest is in the over-all evolution of culture which he sees as a thing *sui generis*. For White, it is not populations which evolve, but culture. This again is a radical departure from the biological theory of evolution. White justifies this departure by calling forth the "superorganic" of Spencer and he readily admits that social evolution is indebted more to Spencer than to Darwin. The superorganic follows its own rules since culture is an extrasomatic adaptation.

It must be admitted that White is correct in one respect. If we consider that man is a single species, and that recent historical developments are merging man's total environment into a single niche, then it is true that we have a kind of unilineal evolution fulfilling the requirements of the biological theory of evolution. But from the point of view of anthropological analysis and the type of questions which biological evolutionists ask, the single-niche approach becomes relatively trivial. It gives us simple answers to very general questions, but it does not enable us to come to grips with the process of adaptation which has developed in individual human societies. Such an approach would be equivalent to a total neglect of chemical, cytogenetic, and population genetic studies, with a total emphasis on a single aspect of paleontology, that of anagenesis. Anagenetic lines of development are of interest, but they are only one small aspect of the total evolutionary process. In addition, they are unexplainable except in metaphysical terms unless we come to understand the underlying process of evolution. This can only be unraveled by attacking process itself on the microevolutionary level.

It is unfortunate that White does not realize that the fact that culture is an extrasomatic adaptation, a proposition that few would deny, does not demand a total commitment to culture as the only subject matter of anthropology. To consider culture as extrasomatic does not require us to abandon the biological model of evolution, since behavior based on culture must still solve basically biological problems. In the next two chapters I shall attempt to show how the biological model of evolution can be applied meaningfully to anthropological problems. Suffice it to say here that once again it is a matter of what questions we ask the data and what framework we begin with as a focus for research.

Multilinear Evolution and Cultural Ecology

Julian Steward is unique among anthropologists of the forties and fifties in his advocacy of multilinear evolution. Steward was concerned primarily with what he called "cultural ecology" and the development of "levels of sociocultural integration." In his work Steward attempted to pick out specific features of the ecological setting that interact with the social system to produce types of exploitative organization. Thus his major focus was on social organization associated with environmental exploitation and its relation to what he calls the "cultural core." The cultural core, perhaps too flexible a concept, consists of those aspects of culture and social organization that are tied more or less directly to technology.

For Steward, levels of social organization are persistent forms. Many levels can occur together in technologically complex societies.

Of his theory of multilinear evolution Steward has said:

"Multilinear evolution is essentially a methodology based on the assumption that significant regularities in cultural change occur, and it is concerned with the determination of cultural laws. Its method is empirical rather than deductive. It is inevitably concerned with historical reconstruction, but it does not expect that historical data can be classified in universal stages. It is interested in particular cultures, but instead of finding local variations and diversity, troublesome facts which force the frame of reference from the particular to the general, it deals only with those limited parallels of form, function, and sequence which have empirical validity. What is lost in universality will be gained in concreteness and specificity. Multilinear evolution, therefore, has no a priori scheme or laws. It recognizes that the cultural traditions of different areas may be wholly or partly distinctive, and it simply poses the question of whether any genuine or meaningful similarities between certain cultures exist and whether these lend themselves to formulation. These similarities may involve salient features of whole cultures, or they may involve only special features, such as clans, mens' societies, social classes of various kinds, priesthoods, military patterns, and the like."

In one of his most successful papers, "Tappers and trappers: parallel process in acculturation" (1956), written with senior author Robert Murphy, Steward compares similar processes of culture change in two ecologically distinct parts of the world, tropical lowland South America and forest North America. In this paper, microevolution involves a shift in social organization that occurs with a similar shift in economic production and commerce. Thus the ecology here relates to similarities in the cultural approach to nature and not to nature itself. This is certainly a valuable approach to the process of cultural evolution, but because of the authors' interest, it leaves out a detailed analysis of primary interactions between human behavioral systems and specifically environmental variables.

A more specifically bioecological focus on the development of cultural behavior may be seen in the work of several scholars. One of the key papers that initiated this school is "Ecological relationships of ethnic groups in Swat, North Pakistan" (1956), by Fredrik Barth. Barth notes, ". . . the importance of ecological factors for the form and distribution of cultures has usually been analyzed by means of a culture area concept." This is a rather broad range scheme in which sets of culture traits are imbedded in rather crude geographic zones. Barth attempts "a more specific ecological approach to a case study of distribution by utilizing some of the concepts of animal ecology, particularly the concept of a *niche* —the place of a group in the total environment, its relations to resources and competitors."

In this study Barth goes on to demonstrate how, within a single broad geographic area, three different ethnic groups have come to occupy special segments, or niches, of the environment.

While Barth concentrates on three niches exploited by three different ethnic groups, M. Coe and K. V. Flannery, in a paper entitled "Micro environments and Mesoamerican prehistory" (1958), concentrate their attention on the multiple exploitation of a range of niches by two single homogeneous populations that inhabited areas with widely differing natural resources. The two groups were the people of the Tehacán valley in Mexico and those of coastal Guatemala. Parallel adaptive exploitation of a range of environments involving successive seasonal movements were noted for both regions.

Flannery offered a similar analysis of Mesopotamia ("The ecology of early food production in Mesopotamia," 1969). But here, in addition to noting geographical zones of exploitation, the author shows how different ethnic groups (settled in different niches) all contributed to a single food-producing revolution:

"The food-producing revolution in Southwestern Asia is here viewed not as the brilliant invention of one group or the product of a single environmental zone, but as the result of a long process of changing ecological relationships between groups of men (living at varying altitudes and in different environmental settings) and the locally available plants and animals which they had been exploiting on a shifting, seasonal basis. In the course of making available to all groups the natural resources of every environmental zone, man had to remove from their natural contexts a number of hard-grained grasses and several species of ungulates. These species . . . were transported far from the biotopes or 'niches' in which they had been at home. Shielded from natural selection by man, these small breeding populations underwent genetic change in the environment to which they had been transplanted . . ."

In another paper ("Archeological systems theory and early Mesoamerica," 1968) Flannery discusses the transition from food collecting to sedentary agriculture in terms of gradual changes in what he calls "procurement systems." Using both the concepts of negative feedback and positive feedback, Flannery notes that for long periods the basic adaptation was not to an ecological zone but to five critical subsistence categories. These were white-tail deer, cottontail, maguey, tree legumes, prickly pear, and organ cactus. Tools used in the exploitation of these key items were similar in two different environments. Exploitation involved the seasonality of the resource and the "scheduling" of economic activity.

Change dependent upon positive feedback came with the development and spread of agriculture. As Flannery says, "We know very little about the nature of early 'experiments' with plant cultivation, but they probably began simply as an effort to increase the area over which useful plants would grow."

Sometime between 5000 and 2000 B.C., genetic changes took place in both beans and corn. He notes:

"For example, beans (1) became more permeable in water, making it easier to render them edible, and (2) developed limp pods which do not shatter when ripe, thus enabling the Indians to harvest them more successfully. Equally helpful were the changes in maize. . . .

"Starting with what may have been (initially) accidental deviations in the system, a positive feedback network was established which eventually made maize cultivation the most profitable single subsistence activity in Mesoamerica."

Flannery then goes on to discuss the effect maize cultivation had on other procurement systems, separating those areas in which maize could only be grown in the rainy season from areas in which it could be grown all year round.

The approaches of Barth, Coe and Flannery, and Flannery involve three variations on a single theme. Barth stresses ethnic differentiation in relation to specific niches, Coe and Flannery emphasize multiniche exploitation by single ethnic groups, and Flannery attempts to show how different groups exploiting different niches through time make a combined contribution to the development of a more complex and highly evolved system.

Another line of analysis has been followed by Robert M. Adams (*The Rise of Urban Society*, 1969). Attempting to find parallels between the evolution of civilization in Mesoamerica and that in Mesopotamia, Adams employs the multiniche, multigroup model but, in addition, attempts to indicate how certain features of social structure developing in the context of ecological adjustment contribute the major share to the process of change. For Adams, "the independent emergence of stratified politically organized societies based upon a new and more complex division of labor is clearly one of those great transformations which have punctuated the human career rarely, at long intervals."

Adams suggests that the preconditions for stratification exist in conical clans (unilinear kin groups in which certain members are considered closer to the central line because of primogeniture and have greater status and/or greater access to common property than others). Such groups survive differentially and contribute to the development of a fully stratified society in the context of increasing technological complexity and political control.

Thus Adams sees an element of social culture providing the thrust for major evolutionary change. The suggestion here is that we cannot limit our analysis of cultural evolution to changes in technology on the level of material culture. Such nonmaterial inventions as money, rent and interest have all contributed to the development and change of behavioral systems. This does not take the study of cultural evolution out of its ecological context, for what is important is the maintenance of an over-all view in which behavioral changes are seen in the context of particular environmental niches, whether they be changes in material, social, or mental culture.

A group of anthropologists, particularly at the University of Michigan and at Columbia University, have concerned themselves with stable ecological systems and the role of behavioral variables in maintaining a balance between population and the natural environment. More recently two members of this group, Andrew Vayda and Roy Rappaport, have also examined the effect of disruption on these systems and considered the role of positive feedback in cultural change. The most comprehensive ecological study employing the notion of stable systems is Rappaport's *Pigs for the Ancestors* (1968). In this detailed ethnographic monograph, Rappaport attempts to demonstrate how the ritual cycle of pig slaughter can be coupled to demographic changes in both the human and pig populations of a highland New Guinea society.

In a more recent set of papers, not all of them yet published, Rappaport has extended his examination of ritual and the concept of sanctity as they operate as communication systems within and among populations. This work is significant for at least the following reasons: (1) it suggests how ritual performances may operate to spread ecologically adaptive information from group to group within an interacting set of populations; (2) it offers an explanation for the high truth value accorded secular information transmitted along with sacred information on ritual occasions, at which times the participants' acceptance of the sacredness and hence the truth of ceremony validates all transmitted messages; (3) it provides a link between ecological studies and the work of structural anthropologists such as Claude Lévi-Strauss who have been concerned with myth and ceremony as information-bearing devices and as systems of transformation.

Marshall Sahlins and Elman Service have distinguished between general and specific evolution in order to clarify the differences between multilineal and unilinear evolution. The distinction concerns evolution writ large as pursued by Leslie White and the various approaches to multilineal evolution outlined here. In addition Sahlins has offered several interesting analyses of social organization under varying ecological conditions. He has paid particular attention to relations between economic organization and social structure, especially family structure. Morton Fried has published several articles and a book (*The Evolution of Political Society*, 1968) devoted to the analysis of political evolution. The cumulative aspect of evolution has been investigated by Robert Carneiro, and Gertrude Dole has published several papers relating changes in kinship systems to changes in economic and subsistence behavior. Finally I have discussed (in *Adaptation in Cultural Evolution: An Approach to Medical Anthropology*, 1970) the role of disease-related behavior, including hygiene and therapy, in ecology and cultural evolution.

IX
CULTURE AND EVOLUTION

Analogue Models

The idea of evolution has long intrigued anthropologists. As we have seen, analogies have been drawn between the proliferation of species and the proliferation of cultures through time, as well as between the progressive development of complexity in organic systems and the persistent increase in man's ability to master his environment. Other more questionable parallels between biological and cultural processes have been offered as building blocks for the construction of a theory or theories of cultural evolution.

But a theory must explain phenomena. Analogies, however useful as heuristic devices, cannot extend a theory into new territory. This can be accomplished only if the existing measures of variation and existing systemic models can be employed on new data. Furthermore, false or misleading analogies may obscure real relationships where they exist and focus attention on spurious connections which may be intellectually fascinating but scientifically useless.

Analogue models of evolution applied to culture fail because they employ false analogies, ignore proper measurement, and deny continuity between biological and cultural processes, the very thing which one would expect from a more general theory of evolution.

Specific analogies have been claimed for such mechanisms as mutation and innovation, diffusion and interbreeding, or species and culture. In the first two sets we have clear oppositions between traits which are transmitted somatically on the one hand and extrasomatically on the other. In the case of culture and species, we must recognize the difference between a class of behavior limited to a single species, man, and a taxonomic device used to differentiate one class of animals from another.

Separatism and the Superorganic

Beyond the problem of analogies which are misleading or incorrect, most social evolutionists have been faithful to a separatist tradition founded on the observation that a gulf exists between human behavior characterized by culture and the behavior of all other species. A new category, a "superorganic," is then created to account for the development and operation of culture.

It has also been argued that the superorganic is a necessary concept because culture exists "outside people," who are only transient members of any particular social tradition. "Individuals die, new individuals are born, but culture lives on." Thus while Darwin finally had the courage in *The Descent of Man* to place the human species where it belonged, in the animal kingdom, anthropologists have been busy for one hundred years re-erecting a barrier between man and the rest of the animal kingdom.

I would prefer to say that individual human beings belong to, or are members of, biological systems characterized in part by a set of behavioral mechanisms known collectively as culture. I would use the biological concept of population to delimit these systems so that behavioral, physiological, structural, and environmental variables could be analyzed in the same context and on the same level as they contribute to the configuration and operation of a system. If we do this, we no more need recourse to a superorganic concept than does the geneticist when he describes the dynamics of a breeding population. Again, individuals die, new individuals are born, but the population lives on. The system, as studied by geneticists on the one hand, and anthropologists on the other, consists of operationally definable parts and significant relationships between these parts. A breeding population taken as a unit has characteristics which can be described fully only in terms of relationships between variables, and dynamics which are different from those of individual organisms. Furthermore the elimination of the superorganic allows us to locate the source of variation (whether it be biological or cultural) where it belongs, on the level of the individual organism.

The concept of system levels which was introduced in Chapter V should not be considered as a semantic substitute for the superorganic. While each system may have its own rules, these rules apply to features of the system in question, not to a new set of properties associated with some order of complexity. In every case the analysis consists of statements about variables and relationships between them. If this were not the case it would be possible to create abstract categories whose properties would be based on the number of nesting subsystems which were imbedded in them. With the possibility that ecosystems are larger unit systems than populations, one might then attempt to justify the use of a super-superorganic concept. This would of course be absurd.

Species Differences

This is not to say that human beings are indistinguishable from other species. Of course man is different, not only on morphological grounds but on behavioral grounds as well. But significant behavioral gaps also exist between ants and ladybugs, cats and dogs, even between cocker spaniels and basenji hounds (see Scott and Fuller, 1965). For too long we have minimized existing behavioral differences between species other than man, not all of which are genetic, and concentrated on the obvious fact that culture, if not totally limited to humans, is at least quantitatively unique in man. As George Gaylord Simpson (1953) has pointed out, culture is man's major adaptive characteristic; a species' specific adaptation. Culture, like wings, prehensile thumbs, and nesting behavior in both birds and bees, consists of traits which are employed in the process of adaptation. Systems of behavior which employ culture differ significantly from systems of behavior which do not, and so some new terms are necessary to describe certain aspects of those systems within which culture operates as a major but not exclusive part. But this does not mean that a totally new system class must be employed to describe human adaptation any more than we require a new system class for every other species' specific adaptation.

Return to the Biological Model

In spite of the misunderstandings and misapplied analogies, I am convinced that the theory of biological evolution is one which has direct and profound implications for all of anthropology. What is required is the replacement of analogue models of evolution with the biological model, and a redefinition of problems within the context of this real model. Let me turn, therefore, to an analysis of biological evolution.

In Chapter V, I outlined the characteristics of biological systems. The following attributes were suggested as crucial to an understanding of such units:

1. Self-regulation
2. Self-replication
3. Creativity
4. Specificity of adaptation
5. Subsystems
6. Redundancy
7. Distribution
8. Molecular behavior (limited primarily to animal organisms)

What remains now is a discussion of evolution as a general theory to account for the operation of biological systems inclusive of behavior and culture.

The theory of evolution is a theory about *how* species (or biosystems) originate and is not about what species originate. It is a theory about process. For purposes of analysis I shall break the theory down into what I call *process statements, process mechanisms,* and *process-outcome statements.*

Process statements refer to the interaction of variables which occur in two realms: systemic and environmental. Following Ashby (1960) I shall call variables which are part of the system "variables" and those which occur outside the system, but which affect it, "parameters." Process statements can then be said to refer to the interaction of variables and parameters.

Process mechanisms act to maintain stability or introduce new variables into the system which can lead to system transformation. Process mechanisms include mutations and other genetic phenomena as well as nongenetic phenomena.

Process-outcome statements refer to changes which occur transgenerationally as a result of interactions between systems and parameters. The outcome of process can be measured in terms of adaptive value in reference to specific environments. The outcome of a system transformation or changes in parameters can be neutral, adaptive, or nonadaptive (the latter is a change away from maximization).

Process statements and process-outcome statements can both be generalized as statements of universal application. We can say, for example, that given changes in variables or changes in one or more parameters, new systems emerge, some of which have a high probability for increased maximization. In certain cases, particularly when there are drastic changes in parameters, the system can be destroyed. These generalizations constitute a statement of the theory of evolution.

Empirical tests of the theory must occur in specific contexts. This is true because the investigator must eventually specify (1) the state of the system under investigation, (2) the parameters affecting the system, (3) the process mechanisms which are operant, (4) the new variables which occur, if any, and (5) the potential outcome as measured in terms of selective coefficients.

Since the theory of evolution is a theory concerned with process, empirical tests must be related to specific evolutionary episodes. So-called general evolution (Marshall Sahlins and Elman Service, 1960) is related only to generalizations about outcome and has little to do with process.

Evolution is a process through which systems develop and are modified in relation to specific environmental backgrounds. All the theory requires is that there be mechanisms of variation (producing new variables) and mechanisms of continuity (preserving maximization) present in these systems and that these systems be subject to environmental selection. There is no requirement that these mechanisms be specifically biological in nature. If the theory is seen as a theory about process, the distinctions between so-called biological and cultural evolutions disappear. There is only one evolutionary process—*adaptation*—and one measure—comparative population size—in the context of a specific environmental niche.

As we shall see in the next chapter, however, there are situations in which population size must be considered only along with other factors in the analysis of cultural adaptation if we are to achieve a full understanding of process. This is because, while on the subhuman level the conversion of environmental energy goes almost exclusively into the creation of organisms, on the human level energy may be diverted to other uses such as the accumulation of prestige or political power. Strictly speaking, each diversion of energy is a loss of biological efficiency, but no system of cultural behavior could be understood without focusing attention on these diversionary factors. Such factors themselves often contribute, at least in part, to changing the capacity of the environment to support more individuals.

Culture traits are the result of process mechanisms just as much as puppy dogs' tails, and process mechanisms which produce or change culture traits are a subtype of general-process mechanisms. Furthermore, in any adaptive system, cultural and biological factors can each modify behavior and each other, and these modifications can then affect the state of the system or act to transform it into another system.

Cultural Process Mechanisms

Let us not confuse the fact, however, that while the evolutionary process is the same in all species, the operating rules of those process mechanisms which produce culture traits are not the same as those which produce biologically determined somatic or behavioral traits. This is precisely why the analogies between biological and cultural mechanisms fail. To call the process of innovation in culture a mutation is to depart from the logical model of evolution. Mutations are somatic, they are recurrent even when negatively selective, and they become fixed in a population only through genetic trans- mission. Culture traits are not usually invented over and over again in the same population, cer- tainly not randomly, and invention itself may be something more than accidental, if something less than perfectly teleological.

The selective advantage of behavioral systems can be analyzed in terms of competition between populations. The adaptive process can also be in- vestigated by studying the states of defined cul- tural systems. The question to be asked is: "How efficient are these systems in terms of self-regulation under stress, as well as in total reproductive poten- tial?" Populations competing for the same environ- mental niche can be studied comparatively in terms of relative success. Cultural succession as studied by ecologically oriented archaeologists (particu- larly Coe and Flannery, as cited in the last chapter) is one key method of analyzing cultural evolution in the perspective of biological adaptation.

Of great importance for anthropology is the study of process mechanisms. Much less is known of these for man than for the biological systems of other species. Recent attempts by Homer Barnett and Margaret Mead to analyze some of these mechanisms are interesting. Barnett, in his book *Innovation, the Basis of Culture Change* (1953), suggests that culture change, particularly the de- velopment of new tools and techniques, results from a process in which individuals discover anal- ogies between items in different realms, combine known elements in new ways, or use old tools for new purposes. Mead, in her book *Continuities in Cultural Evolution* (1964), investigates the relationship between social placement, psychological factors, and change on the micro level. She accepts macroevolution as a process, but chooses to examine evolutionary processes on the level of individuals and small groups. On this level the characteristics of individuals which are significant from the point of view of evolution can be defined as those properties that are most peculiar to a member of those groups having the greatest leverage either for the continuation of the status quo or for innovation in a specified social unit. Mead goes on to suggest that the structure and ideology of these groups may have a profound effect upon the success or failure of the genius or geniuses within it. Thus the family's beliefs about the genius of its most gifted members can create an atmosphere of acceptance or rejection. If there is only one outstanding member of a group, the absence of any others with whom he or she might communicate may inhibit innovation or its acceptance.

Other anthropologists not usually associated in their publications with cultural evolution have nonetheless contributed material that helps to fill in our understanding of process mechanisms.

Alfred Kroeber in his 1939 study *Cultural and Natural Areas of Native North America* was one of the first anthropologists to investigate relationships between environment and culture. The main thrust of his work was the analysis of historical relationships among cultures, but Kroeber considered environmental adaptation as one line of inquiry. He suggested that each culture is influenced by its subsistence base and that environment plays the role of stabilizer for culture. Environment tends to inhibit new variation not only because it acts as a limiting factor on what is possible but also because once an adaptation has occurred it becomes difficult to change the direction of adaptation. This of course is a statement concerning the role of specialization in adaptation. *Cultural and Natural Areas* also reviews demographic data in relation to environment and means of subsistence. Kroeber notes that coastal residence in most cases is associated with increased density, but that agriculture alone does not increase density. In addition, he points out that a broken environment—one that has many local variations—is associated with the diversification of local cultures.

The work of George Peter Murdock has been concerned primarily with social structure, particularly the analysis of relationships among sets of social variables. Although Murdock considers himself critical of evolutionary theories in anthropology, his work can be put in an evolutionary framework as long as we recognize that his model concerns what I have referred to above as internal adaptation and that it is nonlinear.

Murdock's *Social Structure* (1949) considers a set of hypotheses that predict correlations between such aspects of social structure as residence pattern, household type, and kinship terminology. He suggests that shifts in one pattern will lead to shifts in the other patterns as well, and in an ordered and predictable way. Thus, at least for my purposes, his scheme can be seen as an analysis of internal and cyclical cultural process. It is not difficult to link this kind of analysis to external adaptation by looking for correlations between environmental variables and subsistence activities on the one hand and shifts in social structure on the other.

Murdock has always stressed the nonrepetitive aspect of specific change and no doubt feels most comfortable as an ethnohistorian. His book *Africa* (1959) is written as an ethnohistorical account of culture change in Africa and as a test of the theories presented in *Social Structure*. Murdock admits, however, that similar events occurring in different places and at different times can produce parallel effects; and he accepts the validity of both the historical (particularistic) and scientific (generalizing) approaches to culture change.

Recently structuralism as practiced by Claude Lévi-Strauss and his followers has come into vogue in anthropology. Structuralism borrows from the methods of linguistic analysis and is concerned with sets of variables that operate within the context of definable structures. Thus it attempts to reduce the overwhelming amount of available discrete data to sets of finite relationships. Generally these are built up around oppositions between elements which therefore contrast with one another in significant fashion. Sets of contrasting elements constitute a structure. Structures diffuse from culture to culture. The elements out of which they are constructed change through time and across space, but the set of essential relationships is always maintained. In general, Lévi-Strauss does not look for the meaning of individual symbols in the material he analyzes; thus he cannot be accused of forcing meaning on data. Data are the raw material for analysis and the task of the anthropologist is to demonstrate structural relationships in it.

This is not always easy to do and some of Lévi-Strauss's analyses are more convincing than others. In his major work (*Les Mythologiques,* published in four volumes between 1964 and 1972), Lévi-Strauss traces the structural variations in over six hundred myths collected among a large number of ethnic groups ranging from lowland South America to North America. In this work he shows how the basic set of relationships is maintained even with major shifts in particular content and apparent emphasis. The entire work represents an attempt to construct a grammar of myth with which any one myth in a series can be generated from a known set of rules and any given other myth. Although the task is not completely successful, the analysis presented in *Les Mythologiques* is an impressive demonstration of relationship and transformation.

Lévi-Strauss's method proceeds as follows: (1) find the structure in a set of data (a myth, a work of art, etc.); (2) demonstrate the transformations that occur from set to set of related data (from myth to myth, for example); (3) demonstrate the relationships among different realms of culture by showing that sets of transformations transcend one particular cultural domain (from myth to art, for example).

If continuity always underlies change, if structures are permanent, what has structural analysis to do with evolution and the study of adaptation? First of all, Lévi-Strauss himself admits that the specific content of myth as well as other aspects of culture are all influenced by the environment in which a particular population is imbedded. In addition, he admits that particular transformations may be distorted at least to some degree by necessary ecological accommodations. He has not concerned himself with these culture-environment relationships because his major interest has been to analyze basic structures in order to understand better the operation of culture and the human mind.

But remember, evolution involves continuity as well as change. If structural analysis is correct, it provides us with new insights into the conservative aspect of culture. In addition, it might provide us with some clues as to how certain cultural behaviors that might be maladaptive for a set of individuals as individuals but adaptive for the population at large could be maintained within the cultural vocabulary of such a population.

Lévi-Strauss has said that men don't think myths, but, rather, myths think themselves. Just as unlettered individuals do not know the grammar of their own language, primitive man does not know the grammar of his myths. Myths change through time independently of their tellers, according to structural rules. Now many anthropologists have suggested that myth and ritual are repositories of important cultural information. If this is true and if the information contained in them is useful from the point of view of adaptation, then myths may not only store information but also protect it from ideosyncratic change. And if myths provide prescriptions as well as proscriptions for behavior, such behavior may be maintained in a culture for long periods, subject only to a slow process of selection in which only a limited set of transformations are possible.

Up to the present, structuralists have not concerned themselves with ecology or cultural evolution. In fact, their approach has generally been ahistorical, for they have tended to emphasize the timeless nature of structures. The suggestion made here that structural anthropology might provide evolutionary studies with important leads, particularly in the area of mechanisms of continuity, is strictly in the realm of hypothesis.

For too long, social evolutionists have tended to focus their attention on the cumulative aspects of process outcome. This has transformed them into cultural paleontologists, a worthy profession but limited in its approach to behavioral problems. Attention to this aspect of culture has led them away from the biological model of evolution with its stress on process. The framework of biological evolution provides a context within which the analyst can approach anthropological problems with an open mind. If physiological, psychological, or ecological variables are important for a statement about a particular behavioral system, he need not feel guilty of having abandoned man. After all, we are essentially interested in human behavior and not solely in culture.

American anthropology has always claimed to operate within the context of a four-field approach in which physical anthropology shares emphasis equally with ethnology, linguistics, and archaeology. Physical anthropologists have indeed maintained a strong interest in the effects of culture on physical development, but cultural anthropologists have neglected a major biological theory which can and should be extended to human behavior. Good theories are few and far between in social science. Those which exist should not be discarded or emasculated.

American social scientists, most of whom fall within the positivist-determinist tradition, have tended to form schools based upon their faith in a particular first-cause principle. Some are economic determinists, others psychological determinists, still others cultural determinists. If, however, we consider both somatic and behavioral traits in terms of their adaptive value in relation to specific environments, we may find that some variables in a system are psychologically adaptive, others economically adaptive, and so on. Traits which have adaptive value for a population as a whole are also, it should go without saying, biologically adaptive as this term is defined in evolutionary theory. Hence the biological model provides an umbrella for other deterministic theories.

Finally acceptance of the biological model of evolution brings together two rather disparate interests of American anthropology: ethnohistory and behavioral science. The evolutionary approach provides a meeting ground for these fields in the areas of diachronic studies of adaptation and synchronic studies of system states. Each type of study should provide valuable material for the other, providing research problems are carefully structured. The next chapter is concerned with the empirical applications of the biological model in anthropological research.

X
THE ADAPTIVE MODEL

Evolutionary biologists are well satisfied that the Darwinian model of evolution has been empirically verified. But what of its application to anthropology? While most anthropologists agree that "culture" is adaptive, many leave the question there and pass on to particular problems which have no evolutionary import. This is due, in part, to the fact that these individuals have restricted their interest to things cultural, avoiding problems which link culture to biological and other parameters. Others recognize relationships between biological and cultural variables, but for one reason or another do not approach research problems from the framework of a consistent theory. One can only speculate as to why this is true, but I suspect that most anthropologists view evolutionary problems and their solutions as strings of cumulative sequences. Certainly those who have opted the field and labeled themselves as social or cultural evolutionists, be they unilineal or multilineal, have concentrated on this aspect of evolutionary outcome. Their view of evolution is diachronic and sequential. But one can pursue evolutionary problems synchronically in terms of systems analysis seeking to construct models which account for the operation of homeostatic systems as well as positive feedback in human populations. Because evolution takes place through the modification of existing structures, there need be no prerequisite for increasing complexity or strictly linear development. A so-called "regression" such as the loss or modification of a perceptual mechanism, for example, may lead to new twists and turns. Thus in the evolution of the primates, the sense of smell gives way in a large degree to the development of visual centers, and as the brain and neurocranium increase in size, the snout recedes. Evolution is a process in which organisms respond to environmental pressures rather than to some doctrine of linear development.

This does not mean that evolution is totally non-linear. As I have pointed out in the chapters on genetic mechanisms, adaptive sequences do develop and become modified in the context of specific environments in a relatively linear way. Furthermore, the existence of some structures depends directly on the pre-existence of some others. The search for linear chains, then, is *one* important aspect of evolutionary research, whether this be an analysis of physical or behavioral forms. The "fossil record" of societies is far less complete than that of biological forms, however. Social and mental culture is much more difficult to reconstruct from incomplete evidence than the soft anatomy which surrounded or was surrounded by the skeleton. Archaeology is less revealing of past cultural developments than paleontology is of the morphology of extinct forms. Although we know a good deal about technological development, we know only the barest minimum about the ideas and social systems of past societies.

Just as biologists fill in their sketch of sequential

development through the examination of living forms, anthropologists utilize material from less technologically developed living peoples. In this way they attempt to provide correlations between material culture and social and mental culture. This technique is fairly safe in biology because it is possible to determine when a particular grade of animal first appeared in the phylogenetic sequence. Fossil precursors of living forms may then be followed in the paleontological record right up to the living specimens to be analyzed. The situation for anthropologists is by no means as well ordered. Many of the technologically primitive peoples living today are remnants of populations which have been pushed into marginal territories to be surrounded by more advanced peoples. In many cases these retreats have been marked by a loss or severe modification of cultural traits. Furthermore, the migrations of peoples across ecological boundaries violate the constant of adaptation to relatively stable environmental niches which can be applied to animal species which are linked to their fossil ancestors.

Linear Sequences and Scaling

Some scholars, most notably Robert Carneiro of The American Museum of Natural History, have attempted to construct scales which reflect both logical and empirical sequences of cultural development. Using material from the archaeological record and from living peoples, this method can determine with some success which forms provide the precedent for later developments and modifications of culture. The scale is ordered in such a way that chains of related traits, i.e. traits related to a single dimension of development, let us say grain agriculture, can be built. Thus if we find item E on the scale in a culture, we can be fairly confident that items A, B, C, and D will also occur. Conversely, if we find items A and B, but not C, D, or E, then we will not expect to find F, G, H, or any others further along the scale. This technique was originally developed by L. Gutman and employed by E. S. Bogardus to see if attitudes towards minority groups could be ordered. Bogardus assumed that people willing to marry members of a minority group would also be willing to have them as neighbors, as members of their clubs, as fellow workers, as residents in their country, and as visitors to their country. Each item on the scale from visitor to marriage narrows the social distance between the respondent and the hypothetical member of the minority group. If the scale is ordered, then a person who would not want a person to be a member of his social club would also not want that person to marry into his family. Or abstractly, if items A and B only are checked, no other items on the scale will be checked, and if item G is checked, then all preceding items will also be checked.

Scale analysis in anthropology can tell us which traits follow an orderly sequence of development and which traits do not. In addition, it provides a check on the relatedness of traits to one another as members of a single dimension which can be sorted out of a mass of cultural material. It can, for example, tell us that a particular form of social behavior occupies a consistent place in the sequential development of some technological system. If such a behavior fits into the ordered scale, then it is legitimate to assume that it occupies a necessary place in the development of that linear sequence, and that it is a part of that sequence.

Scale analysis, as it is currently pursued, is unable to relate adaptive processes to environmental parameters, nor can it tell us much about the presumably large number of social forms which will not scale, but which nonetheless change as systems change. Furthermore, it can tell us nothing of the mechanisms which underlie change, and thus has no application to the analysis of dynamic self-regulating systems. In sum, its function is primarily paleontological rather than genetic.

Tests and Hypotheses

If the biological model of evolution presented in this book has validity for anthropology, the model itself should suggest means of testing its propositions. The most obvious idea would be to look for situations of competition between populations. This certainly can be done, and I shall present some research strategies for this approach below, but this approach cannot answer all the questions which must be posed in the analysis of human behavioral systems. Since culture traits are not bound to the germ plasm of specific individuals or groups (the fact that they can diffuse), the picture of culture as an adaptive mechanism must include borrowing. The issue is further complicated by effect of the quasi-teleological nature of culture on positive feedback or maximization. Culture traits can be invented through conscious effort, and although such traits can affect a behavioral system in unsuspected ways, man does have a good deal of control over his own destiny. The quasi-teleological nature of human adaptation must be worked into the theoretical model of evolution as far as human behavior is concerned. While this problem provides further complications for the model, it does not destroy it, since over-all maximization is still the major criterion for successful adaptation, and the measure of over-all maximization is the same one applied to all other types of biological systems.

Population Competition

Let us first consider competitive populations. Certainly adequate evidence exists that human populations have often been replaced through direct confrontation of competing groups. Europeans have almost completely replaced the American Indian in the New World, the Tasmanians are extinct as a result of their inability to compete with European invaders, and Negro Africans have almost completely replaced Bushmanoid peoples in East and South Africa. These are all examples of large-scale migrations and replacements, but it is just as likely that small-scale operations of the same type have occurred again and again between human populations competing for space.

Teleological and Nonteleological Adaptation

When populations are in direct competition, the selective advantage which one has over the other may be easily perceived either by those directly concerned or by the anthropologist. Thus, to take a simple case, where two tribes competing for space are made unequal by possession by one group of bows and arrows, the military leaders of the other group may not only recognize their disadvantage, but attempt to do something about it! They may send a war party out to capture this "secret weapon." If they are successful, the military hardware is soon equalized. This, in effect, has been the history of warfare. Obvious military advantages are soon overcome through the conscious attempt on the part of the underdogs to adopt the superior weaponry of their adversaries. But let us assume that some other factors contribute to an inequality between competing peoples—an inequality which confers a selective advantage on one, but which is less obvious. Let us assume, in the case of competing tribes A and B, that both peoples practice cannibalism. Let us further assume that group A eats people rare and group B eats people well done. Now if there was some disease present in the populations which was spread only through the ingestion of undercooked human meat, group B would have an advantage over group A. Members of A are disadvantaged through a cultural practice which is unrelated to the pattern of warfare. The same kind of situation might occur in areas where actual confrontation through warfare may be only peripheral to the fact of increasing population on the one hand and a receding population on the other. If a group of tribes within an area fight each other and eat undercooked human flesh, the entire set of populations may be reduced to the point where some outside peoples could move into and take over their territory. I will admit that the example used here sounds farfetched, but I have chosen it because it is very close to a real situation.

A disease has recently been discovered among a group of New Guinea highland natives which might very well fit the pattern described. The disease, known as kuru, affects the central nervous system and is almost always fatal. Its occurrence is restricted to a small number of related tribes in the Foré region of highland New Guinea. Children of both sexes are affected, as well as adult females. The occurrence of the disease in the population is frequent enough to threaten the very existence of the populations carrying it. For a long time it was assumed that kuru was genetic in origin because it occurred only in members of the Foré population, and might occur even in Foré people who had migrated away from the region. If the disease was infectious, it was reasoned, other peoples would get it, and conversely, migrant Foré would not develop symptoms when they were out of contact with infectious centers of the disease. If the disease was genetic, on the other hand, its high frequency had to be explained. Such a frequency could be based on either genetic drift or some kind of adaptive polymorphism in which the gene for kuru conferred an advantage on the population as a whole. It has been suggested that the outbreak of the disease might be related to some new cultural practice which destroyed the advantage conferred by the gene by changing its penetrance. This in itself would confirm the point which I am trying to make in this chapter, since a change in behavior is posited to affect the operation of a gene in a population with harmful consequences. Thus behavior of a nongenetic sort would be implicated in the effect of a gene and also indirectly with regard to the selective advantage of that population *vis-à-vis* other competing populations.

There is strong evidence at the present time, however, that kuru is infectious rather than genetic. Gail Williams et al. suggested that a look at the cultural practices of the Foré might present some clues as to the etiology of the disease. Among the practices which they sorted out of the cultural material was the fact that the Foré people eat the brains of their victims. They also found that residence patterns are such that women and children have greater access to undercooked food than do men, since women and their young children of both sexes live together away from the men. Women are responsible for food preparation, and it was suggested they and their children might taste undercooked food which might never get to the men.

If kuru is an infectious disease of the central nervous system, the infectious agent might well be restricted to the human brain. Furthermore, well-cooked brains would not be dangerous since the agent would be destroyed in the cooking process. Women and children of both sexes would then be exposed to the disease, and men would be protected from it. The effect on the total population would nonetheless be drastic since adult women would die of kuru before they had produced their quota of children. In some areas of the Foré region the death rate from kuru alone is responsible for a significant decrease in the population. But what about the fact that Foré people who had left the area and their old practices to work on European plantations still came down with kuru? This fact could be explained if the disease had a long incubation period. As a matter of fact, there is a disease of sheep and goats, scrapie, which has an incubation period of from eighteen months to three years and which affects the central nervous system. Kuru could well be a disease of this type. Thus the infectious focus of kuru in the Foré region might result from a combination of behavioral factors which limit the spread of the disease plus certain genetic factors which predispose the population to infection. Indeed, the infectious nature of the disease has recently received confirmatory evidence. Dr. Carlton Gajdusek of the National Institute of Nervous and Neurological Disorders has successfully transferred kuru from human cerebral material to apes!

Adaptation Within Population Groups

I should now like to review a series of studies and hypotheses which relate behavioral traits to biological adaptation. The distributions of these traits in human populations would be explained on the basis of their adaptive significance. Such adaptations can, no doubt, develop as the result of positive feedback within populations. Population displacement may be, but need not be, a factor in their spread and maintenance.

The Harvard anthropologist John Whiting has been concerned for some time with cultural practices related to birth, weaning, and puberty. Over the past several years he has demonstrated a series of correlations between types of sleeping arrangements, postpartum sexual taboos, residence patterns, puberty ceremonies, and polygyny. Correlational studies are useful in that they demonstrate relationships between variables, but they are usually unable to provide a causal explanation of phenomena. If two items, A and B, are correlated it is impossible to demonstrate which is the independent and which is the dependent variable. In an attempt to solve this problem, Whiting searched for an environmental variable which could lie at the base of a correlative sequence. Whiting reasoned that if he could show a relationship between some aspect of the natural environment and behavior, it would be more logical to assume that the environment had caused the behavior than the behavior the environment. He found that long postpartum sexual taboos and late weaning occurred in high frequency in tropical areas and that conversely short postpartum taboos and early weaning were common practices in temperate areas. Since he could find no direct explanation for such a relationship, Whiting searched for a hypothesis to account for the geographical distribution of the traits. Tropical areas of the world are also areas in which the dietary balance of human populations is generally high in carbohydrates and low in protein.

Whiting's hypothesis stated that long postpartum sexual taboos on mothers and consequent late weaning of infants were correlated with a low-protein diet. The reasoning behind the hypothesis was that late weaning might offer infants protection against kwashiorkor, a severe form of protein malnutrition.

Postpartum sex taboos prevent pregnancy in women while they are nursing an offspring and this practice normally prolongs the nursing period. In most societies, when a woman becomes pregnant she discontinues nursing shortly thereafter. In many cases this custom is justified by the belief that pregnancy spoils the mother's milk.

Interestingly the general correlation between long postpartum taboos and tropical climate did not hold up in tropical South America, although late weaning is also common there. This forced the researchers to search for an explanation of the exceptional case which did not violate the original hypothesis. It was found that induced abortion is a common practice in those areas. Such a practice has the same function of spacing live births and preventing nursing during pregnancy as a long postpartum taboo. In this instance the exceptional case actually strengthened the hypothesis that the geographical correlation is related to late weaning rather than to postpartum taboos themselves, and that the relationship reflects a biologically adaptive behavioral pattern.

An interesting feature of Whiting's study and one which must be dealt with here is the suggestion that a basic adaptive relationship may tend to shape other behavioral patterns through a causal chain. Whiting's original study was actually concerned with a high correlation between sleeping arrangements in which the mother and child sleep together without the father, long postpartum taboos, and the incidence of circumcision. The relationship that finally emerges between tropical climates and circumcision is based on intervening variables, beginning with behavioral adaptation to low-protein diets. Thus, circumcision may occur in tropical climates and yet not itself reflect a specific biological adaptation to such a climate. This custom, it is suggested, is an outcome of a basic adaptive relationship plus certain psychological factors which themselves influence human behavior. The correlation between circumcision, sleeping arrangements, and postpartum taboos, as well as a series of other culture traits investigated by Whiting, reflect a complex adjustment to human biological, social, and psychological needs. Such adjustments should tend to maximize the whole behavioral system of populations and thus reflect biologically adaptive adjustments.

It appears, from a casual perusal of the literature, that the use of human fecal material as fertilizer is restricted to geographical areas which reflect high population density and hence strong pressures on the land. Soil fertility and productivity are threatened in such areas unless all available fertilizers are utilized. (In technologically and scientifically underdeveloped areas substitutes, such as naturally occurring phosphate deposits, must be considered as culturally unavailable because their potential use is not understood.) But why should human fecal material not be used in other areas as well? After all, it would increase soil fertility, as well as crop yield, and therefore cut down on the amount of labor necessary to produce adequate food for the population. The answer might well be that in areas where population density does not threaten productivity, the dangers through health hazards of this practice far outweigh any benefits which might accrue from it. Human feces carry a large variety and number of infective agents which are dangerous to human health. I would suggest therefore that strategies have developed through time which tend to reject such behavior patterns in areas where their harmful effects would outweigh their benefits. Such patterns develop conversely where they tend to maximize survival for the populations involved.

A similar case may well apply to the eating of domestic animals which have died from natural causes. In this respect, societies can be divided into two types: those which taboo such animals and those which include them as accepted items of diet. The harmful effects of such a practice relate to the transmission of disease from animals to man. The benefit comes from an addition of protein to the diet. If our adaptive model works in this case, we should expect to find carrion eaters in areas where protein is scarce and in which it is difficult to maintain large herds of domestic animals. Carrion would not be eaten where animal protein is relatively abundant, and where domestic animals may be raised successfully in adequate numbers. There is a further complication here, however. Cattle of various sorts may provide high-quality protein for consumption through milking. The hypothesis would have to be restated thus: In areas where animal husbandry is difficult, where yields are low, and where domestic animals are not milked, domestic animals which have died a natural death will be eaten. Conversely, in areas where domestic animals or other protein sources are plentiful or where milk is exploited as the protein source carrion will be avoided. Thus in West Africa in areas where cattle are only one source of animal protein and where the rainfall is plentiful to support good grazing land, carrion is taboo. In the savannah-desert littoral, on the other hand, where cattle raising is more difficult and where fish are generally not available, carrion is eaten. Islamic herders, however, exploit their animals for milk and taboo the eating of carrion. A major exception to this hypothesis occurs in East Africa, where cattle people milk their animals but also eat them when they die of disease or old age. In this particular case the populations are dependent to a large extent on cereal food which is obtained through the trading of cattle with agricultural people. Milk and blood drawn from live animals do not supply adequate protein. Live cattle represent the storage of convertible food resources and act as a hedge against bad years. Large herds are important to survival, and carrion is an economically safe and nutritionally necessary route to protein.

Minimax

The concept that I am trying to develop here is that of *minimax,* a term I borrow from game theory. The term refers to strategies in which the attainment of net gain is maximized through attention to potential loss as well as potential gain. A good strategy in game theory is one in which a player protects himself against excessive loss as he attempts to win at the expense of his adversary. Thus if a choice exists between winning eight points but risking five and winning six points but risking only two, the correct strategy would be to choose the latter play. If human behavior involves adaptive mechanisms, then one would expect successful strategies to develop through time which minimax behavior in relation to specific environments. The two strategies discussed above, the use of human fecal material as fertilizer and the eating of domestic animal carrion, should occur only in those areas in which the possible loss to the population is balanced by the gain. The gain itself can be measured only in terms of specific environmental settings. The distribution of traits such as those mentioned can be investigated in terms of those variables which would affect the minimax outcome.

It should be possible to test the values and the distribution of such strategies against a hypothetical situation through the use of computers. By feeding the computer information on the positive and negative effects of strategies as well as information of population density, soil fertility, availability of food resources, etc., we could ask the computer to tell us where certain behavioral patterns would be most and least likely to occur. If the distributions predicted by the computer matched the actual geographic distributions of such cultural traits, this would tend to confirm the hypothesis that human groups do in fact develop minimax strategies in relation to environmental parameters.

At this point let me introduce a more complicated hypothesis. It is well known that pregnancy and the first year of life are surrounded by a varying series of proscriptions and prescriptions. If these are examined in terms of their medical value, some appear to be adaptive and others nonadaptive. The postpartum taboos on sexual relations studied by

Whiting are one case in which the proscription appears to have an adaptive value in a specific setting. I would suggest that such behavior patterns are more likely to be adaptive in the medical sense in small populations than in large populations. Selection pressures could be relaxed somewhat in larger populations without threatening their survival and because other intervening variables could then be introduced into the adaptive model. That is to say, small populations face a direct biological challenge to survival, while larger populations may minimax strategies surrounding birth and infancy in terms of variables other than those which relate so directly to the support of population numbers. Now there should also be cases in which the actual size of the population would influence the outcome in the opposite direction. If a relatively small population is limited in terms of exploitable space and resources, then one might assume that behaviors associated with birth might operate to maintain a small stable population. Small island populations would be the ones most likely to exhibit this type of behavior. Thus I would submit the corrected hypothesis that where resources are limited, the sum of behaviors associated with pregnancy and the newborn will not be as medically adaptive as in those situations in which available resources and space are plentiful. The "true" adaptation would be population maintenance. If the hypothesis was borne out, I would still have to explain why less medically adaptive behaviors occur in large populations. Here I would suggest that as absolute survival of the population becomes less of a problem, that new factors come into play which affect the choice of strategies. As I have mentioned above, these may be related to other aspects of system maximization. Restrictions on women then might be related to the maintenance of status groupings in the society, or act to reinforce some religious belief which in turn might be related to psychological or economic factors.

The point I am trying to make here is that as societies grow more complex and as the confrontation between the environment and the population becomes less acute in terms of meeting *basic* biological requirements, the variables which operate in the formation of effective over-all strategies grow in number and complexity. This may be one reason why social disorganization is so frequently encountered in complex societies. Strategies become so complex and alternatives so numerous that evolving systems come to require tremendously long periods for various feedback networks to come into effective balance with the environment. This creates a paradox of its own in that the rapidity of culture change keeps behavioral outcomes constantly off balance. These outcomes are unbalanced further by radical changes in the environment, themselves brought about by the results of human behavior. The analysis of such systems by the anthropologist requires the development of a minimax matrix in which the effects of behavioral choices are studied in relation to a wide range of variables and potential outcomes. In order to do this, it becomes necessary to tease subsystems out of the total system for analysis. In even the simplest societies the number of potential variables is almost infinite. The study of system maximization or system regulation can be simplified by analyzing subsystems constructed of a number of variables related to specific aspects of adaptation. Thus in the living organism one may choose to concentrate on thermoregulation alone. In social analysis, research may be concentrated on those social relationships which operate in the maintenance of a balanced distribution of goods. If a self-regulating system can be demonstrated, and the variables which operate to keep the system in equilibrium can be defined, the investigator will have gone a long way in demonstrating one aspect of the adaptive model of behavior. When similar systems occur in similar environments, the case for causal relationships between environmental parameters and behavioral responses is strengthened. A comparison of self-regulating systems in a range of societies living in similar circumstances might well reveal relative degrees of adaptation which can be measured in terms of the amount of stress such systems can accommodate without breaking down.

A Self-Regulating System

Let me digress at this point to give a brief example of a rather complicated self-regulating system which operated until recently in a population in Bihar, North India. The village, as described to me by a medically trained informant, can be divided into various environmental zones such as dwelling space, rice-paddy land, pasturage (mixed forest and savannah), and forest. The total distribution of available land resources is restricted in the north by the Nepalese border and on all other sides by the holdings of other villages. Most of the villagers are farmers, who cultivate rice and raise cattle, buffalo, and smaller domestic animals. Many villagers own their own land, but some work for wealthy landlords who pay for agricultural labor with an established part of the crop yield. Until recently paddy land was extensive enough so that savannah and forest were left relatively untouched by the population. Considerable tracts of forest were preserved by the large landholders as hunting territories. Cattle were set out to graze on grass and mixed forest land in the daytime. At night game animals such as deer and antelope moved out of the forest and grazed the same areas. The more dangerous wild animals, quite prevalent in the area, kept well into the forest where game was plentiful. The total human population was stabilized by malaria, which occurred in high frequency and which produced a high mortality rate. In short, a rather good balance existed between the population and available resources. There was no degradation of the land, and the various environmental zones which contributed to the well-being of the village maintained their integrity. After independence, an antimalaria campaign was introduced. The result was a sharp increase in population. At the same time, an increase in contact with the outside stimulated the peasants to sell their crops and use the money for new consumer goods. The amount of available rice in the community rapidly decreased. This in turn brought pressure on the land. Paddy was increased to the point where it encroached on both grassland and forest. Cattle had to go farther and farther from the village to graze, and the buffer zone between the wild predators and cattle was reduced. The predators began to prey upon domestic animals. The

number of cattle was further reduced as pasture-
land became scarce. The protein diet, which consisted primarily of milk products and meat, was
reduced to the point where the government had to
introduce a powdered-milk relief program. In a
short time a population which had been almost completely self-sufficient and which lived in balance
with its environmental resources became dependent
upon government support. The eradication of malaria and the introduction of a new series of consumer goods were both responsible for this change
away from a stable system.

Empirical Demonstration of Self-Regulation

In order to demonstrate that this was originally a
self-regulating system, I would have to show that
certain variables in the system changed their values
in an orderly way in response to pressures on the
system. In this particular case I would suspect that
the incidence of malaria may have been dependent
upon population density and hence a variable in
the system. In addition, big-game hunting probably
regulated the wild-animal population. There must
have been some mechanism within this system
which operated to maintain a reasonable balance
between the human and the domestic animal populations as well. Such a mechanism has been indicated for natives of the New Guinea highlands by
Roy Rappaport of the University of Michigan. Professor Rappaport has demonstrated that cycles of
religious ritual involving pig feasts act to regulate
the demographic distribution of both men and pigs.

It may be that in the Bihari village as well ritual has a similar effect on the system.

A scientific study of the Bihari village in relation to an adaptational model would require quantification of those variables which appear to maintain the system. I have offered rather impressionistic data here merely to illustrate the kind of study which would be fruitful for analyses of adaptive systems.

One of the interesting apparent features of the Bihari system is the maintenance of hunting territories for the rich. It appears as if this practice had an important effect on the total well-being of the community. These preserves helped to provide a buffer zone between domestic animals and wild carnivores. This is but a minor illustration of the fact that aspects of behavior which may appear to be dysfunctional or merely unrelated to an adaptive system may play an important role in establishing and maintaining a balance. Another interesting illustration of this point has been provided by Marvin Harris of Columbia University. Harris examined the role of sacred cattle in Indian society and found that far from being a burden to Indian society as many suppose, the cow is a major factor in the survival of the population. In addition to providing protein either as meat (to non-Hindus) or through milk production, the cow is one major source of traction, and also provides the most commonly used fuel, dung cakes.

A priori assumptions about what is and what is not adaptive in cultural systems must be eliminated if we are to have successful research designs concerned with the evolution of human behavior. It must also be made clear that what must be searched for in systems analysis (which is only one part of evolutionary studies) is a series of linked variables which operate in feedback relationships to maintain some state of the system. This means that functional statements about such systems do not refer to cause. The origin of traits in a system is a question quite apart from the operation of that system.

Mechanisms of Continuity and Variation

Let me return now to an analysis of the adaptive process. As I have stressed above, the basic relationship between man's somatic structure and culture is a permissive one. Man is, behaviorally, the most flexible of all animals. Humans are born with a largely uncoded behavioral potential, the capacity to learn language and culture. The specific forms which culture takes should therefore reflect the effects of environmental parameters on the developing child. These parameters include aspects of tradition (culture) which are transmitted from other humans (adults as well as peers) and noncultural environmental pressures. This open aspect of human nature has been stressed again and again by anthropologists. It appears, however, that other biological mechanisms acting alone and in combination with learning are also important as mechanisms of continuity and variation in the development and maintenance of human behavioral systems.

Learning Fixation

The maintenance of continuity is enhanced by a biological mechanism which has been somewhat neglected by anthropologists. I shall call this mechanism *learning fixation*. Linguists have thoroughly documented the fact that the ability to mimic the sounds of a language or dialect decreases with age. It would appear that after initial learning has taken place, it becomes difficult for substitutions to be made in the basic code structure. (This does not mean that new items cannot be added to an existing code structure once it has been mastered.)

Edward Bruner, an anthropologist, has suggested that what is learned early in life tends to be retained tenaciously, while items learned in adulthood can be modified or rejected more readily. Of course the entire basis of Freudian psychology rests on the assumption that early experience has a profound and lasting effect on later behavior. Furthermore, our knowledge of human physiology tends to support the hypothesis that this is a biological property of man. Human infants are born with incomplete and immature neural structures. This is a mechanical reflection of the open system which lies ready to be coded. The infant, because it is an incomplete and helpless creature, must be brought up in a social context. This in turn ensures that a good part of its learning will be patterned according to tradition. It has been fairly well established that abnormal parents tend to socialize their children in an abnormal way. The end product of aberrant learning is aberrant behavior.

From all this evidence it would appear reasonable to suggest that as the neural structures mature some of the inherent flexibility is lost. The partial attrition of this flexibility would tend to support cultural norms and hence reinforce the continuity of the cultural system.

Game Playing

On the other hand, mechanisms which produce behavioral variation are also a part of man's biological heritage. These surely provide some of the basis for culture change. One of these mechanisms is a general property of mammals which has been retained in man. I refer here to the play element. Game playing, as Omar K. Moore has pointed out, is a major aspect of pattern learning. Children acquire large chunks of their culture through games in which winning is not crucial for survival. Losing is not a severe punishment, but winning is a rather strong reward. Games are also a nonthreatening outlet for creative experimentation. The creative chess player is able to maximize his winning strategies. Beyond this, artistic production is also a form of game playing in which the artist experiments with form, techniques, and materials. Such experiments may well lead to innovations which ramify beyond art to technological advances. This play element, combined in man with high intelligence and a good memory, provides a powerful source of potential variation.

Many authors have noted a connection between science and art. The scientist plays with a set of concepts and discovers new combinations of elements. In our own form of society this aspect of science is protected through institutional guarantees. Ideally, no demands are placed on the pure scientist. He is allowed to play the game, although the economic structure imposes certain restraints and winners are favored over losers. Normally the scientist is not punished for exploring what may be nonpractical interests—yet the results of such exploration are often major elements in culture change.

Population Pressure

Ester Boserup in her book *The Conditions of Agricultural Growth* and M. J. Harner in a paper entitled "Population pressure and the social evolution of agriculturists" have suggested that population pressure may itself act as a stimulus to change. As a population increases, it approaches the limits of productivity of its environment in relation to its specific technology. As people begin to suffer from nutritional deprivation, they may be stimulated towards search behavior leading to an amelioration of conditions. When the surrounding environment is empty and when such new space can be worked with no changes in existing technology, the most likely response will be expansion within the niche. If, on the other hand, the niche is filled, then it is likely that some sort of behavioral change will occur. This may include (1) technological innovation (or borrowing) involving the development of new tools, techniques, and labor intensification, which raises the carrying capacity of the environment; (2) the introduction of, or increase in, warfare and raiding to capture goods or territory; (3) an increase in, or the introduction of, trade resulting in an increase in food for the population.

If innovations of the type suggested above are not successful, the population may either begin to degrade its environment, a process that must eventually lead to a decrease in population, or introduce methods of population control. The latter may include various birth-control techniques, infanticide, geronticide (the killing of the old), or abortion. The most common types of birth control found in nonliterate society are postpartum taboos on sexual intercourse for the woman, usually from several months to one or two years, marked by such factors as the weaning of a child or the taking of first steps; and coitus interruptus.

It is important to make a distinction here between the original state of cultural development (first domestication of plants and animals, first intensification, etc.) and conditions for change in which a particular population is imbedded in an environment including other populations already practicing different forms of technology. In the first case, technological change must involve innovation; in the second case, new techniques may be borrowed from surrounding peoples. We know that among populations today and in the historical past in different parts of the world change is often stimulated by demographic pressure and merely requires the adoption of techniques practiced by neighboring peoples. Thus slash-and-burn agriculturists who require long fallow periods and thus a great deal of land may under the proper conditions settle down to intensive agriculture with the use of fertilizers, a system of production that requires more labor but much less land. The adoption of agriculture by hunting-and-gathering peoples in the recent past is another case in which borrowing was much more common than innovation.

We must not lose sight of the possibility that under pristine conditions population pressure was not always the stimulus for change. There is a good deal of evidence from archaeology that the domestication of plants and animals was a gradual, and in many ways an accidental, process in which certain plants and animals occurring in close proximity to human populations slowly changed and were changed by human effort. Domestication apparently developed as a convenience that rendered life somewhat easier and less subject to the vicissitudes of natural variations in abundance. Agriculture and animal domestication first developed in the context of very specific environments in which, because of natural conditions, uncertainty and the amount of labor required for subsistence were both reduced. This is quite different from most cases of secondary development in which diffusion played the major role in technological change. There are, or at least were in the recent past, many areas of the world in which hunting and gathering was a highly successful mode of subsistence, requiring relatively little effort under conditions of low population density. The same thing holds true for slash-and-burn agriculture, a method which under conditions of abundant land is highly productive and requires little labor. We have cases, which occurred not too long ago, in which hunting-and-gathering populations were reluctant to adopt agriculture forced upon them by government "experts" because it was, in their eyes, an unnecessary burden.

One further word of caution is necessary before leaving the topic of population pressure. Population is likely to be the major force driving change when the groups concerned are egalitarian in structure and are not subject to outside pressure. The development of status differentiation as well as colonialism created situations in which increased productivity was forced upon populations, or segments of them, for the benefit of those in power. Under these conditions, population pressure may have little or nothing to do with changes in technology. Furthermore, demographic change itself may respond to labor demands created under some system of exploitation. If the natural resources and agricultural production of a population are exploited by outside interests who have no desire to invest much capital in their enterprise, labor intensification is the natural outcome. In situations of this sort, the exploited group may respond to economic necessity by increasing their fertility. This may be done by abandoning methods of population control and by decreasing the age of marriage for females. There is some evidence that this is what occurred in parts of Java where wet-rice agriculture is practiced. It also appears that this is what happened in Ireland in the centuries preceding the potato famine. In these cases, increased population can be an adaptive response to conditions imposed from the outside. From the long-range ecological point of view the results may be disastrous. This is, of course what happened in Ireland. English

colonial interests created a situation in which large
quantities of meat and grain were grown and ex-
ported to the mother country. As more and more
land came under appropriation, the Irish peasant
was forced to use marginal land for his own subsistence. The introduction of the potato at first provided the basis for a demographic expansion. The
potato grows in poor soil and produces high yields
with low labor. Potato gardens could be maintained
by children and, in their spare time, by adults. The
major labor effort therefore could go into the plan-
tation agriculture of the English absentee landlords.
At the same time, the average age at marriage for
Irish peasant females dropped. This increased the
average fertility period per female and decreased
the time between generations. As the population increased, more and more marginal land was put into
potato production. Finally the blight, which came
from Europe, struck Ireland more severely than on
the Continent. Widespread famine was the result.
Many Irish emigrated to England and America
thus swelling the labor force of those countries. Interestingly, shortly after the potato famine, age at
marriage in Ireland went way up.

At the present time, anthropologists do not know
much about how populations respond to economic
pressures. Nor do we know how long it takes for
changes in demographic rates to occur in response
to these pressures, if indeed they do change in more
or less regular fashion. With all the current discussion of "zero population growth," it would seem
important that more attention be focused on these
particular problems.

The Basis of Adaptive Change and Continuity

The kind of change I have been discussing here does not necessarily depend upon the replacement of populations through competition. It is teleological in the sense that many new adaptive techniques *are* sought consciously: nonteleological in the sense that the total effect of planned change on the system may be unanticipated. Such unanticipated consequences of change may bring about an entire shift in the system.

The learning element in animal behavior is based upon drive reduction in which successful behavior is rewarded. The simple classical model consists of *drive states* (a physiological imbalance in homeostasis) and a *cue* which stimulates a *response*. If the response is appropriate, the actor is *rewarded* and the drive reduced: homeostasis is restored. Basic or primary drives are biologically determined and diffuse. Hunger, thirst, and sexual desires are examples of primary drives. These can be modified through the learning process to produce, in man, culturally patterned secondary drives. The desire for an ice cream sundae is an example of a culturally patterned secondary drive.

For animals that learn, the environment is a constant source of reward and punishment which acts upon behavioral variation in such a way that random responses become shaped into adaptive patterns. The transactions between these animals and the environment are of two types: individual learning from experience and natural selection for forms of behavior which are adaptive. The fact that learning is an element in adaptation is significant for all animals. As far as man is concerned, the ability to pass great sums of accumulated knowledge on to new generations through cultural mechanisms makes learning a vastly more powerful adaptive weapon. If we accept the fact that behavioral variations are rooted in the individual and that the environment acts on these selectively, we can eliminate any tendency to view the adaptive process as a kind of environmental determinism. Evolution is opportunistic. Natural selection does not create anything; it merely selects out certain available forms from an array of variation. That parallels exist between different groups in similar environments means that some, but by no means all, behavioral (or somatic variations, for that matter) are repetitive and that the selective pressures which have acted on them are nearly equivalent.

In addition to this conscious, quasi-teleological change, behavioral variations can become fixed in cultural systems through an unconscious process. That is, minimax strategies can develop as unconscious but repetitive outcomes of behavior.

Behavioral choices increase in number and complexity with the development of more complex cultural systems. A matrix of rewarding responses can develop in such systems as part of an unconscious process in which anxiety levels are regulated through a series of culturally patterned responses. Variations in the established pattern can on occasion have a reward value beyond existing behaviors and thus become fixed in the system. On the other hand, changes in an established pattern can themselves be anxiety-provoking because they reduce the predictive value of culturally determined responses. The tendency toward conservatism therefore has a high positive value in a balanced system because individuals tend to constrain one another in an effort to relieve personal anxiety. In many ways culture operates as a reflexive process of mutual constraint. Again, depending on the circumstances, the process mechanisms can operate either to generate change or to maintain continuity in the system. In this case, a new behavior must have a positive value greater than the negative value which any change is likely to produce in a balanced system.

Furthermore, while cultural patterns are generally anxiety-reducing, certain of them are undoubtedly positive or adaptive for the population as a system unit and yet maladaptive for some of the individuals who constitute a part of that system. Culture does not make all of the people happy all of the time. This returns us to the problem of adaptation as a biological process. Psychological factors alone cannot explain the maintenance of certain behavioral patterns which characterize a total system on the population or ecosystem level, because the relationships which exist between these patterns may follow rules which apply not to individuals but to the system as a whole. Thus I would disagree with Tylor that social action is the mere sum of individual actions. It is possible to predict that certain correlations will exist among various components of culturally patterned behaviors, or that a change in one pattern of behavior will lead to an orderly change of some other pattern, without recourse to psychological rules.

It is also possible to study certain aspects of behavioral systems without considering environmental parameters. I have already pointed out that adaptation is both an inward- and outward-directed process. What this means is that the maintenance of homeostasis depends upon goodness of fit among constituent units of the system (inward-directed) as well as accommodations to environmental pressures (outward-directed). An anthropologist who studies the relation between one cultural variable and another is concerned with inward-directed adaptations rather than adaptations to some environmental factors. This is a legitimate and important area of research and has, in fact, been the major traditional orientation of anthropology. Such studies, however, can yield only an incomplete understanding of the adaptive process.

Culture is a code system just as genetic structure is a code system. Codes are information-transferring devices. The genetic code is maintained in the DNA molecule; the cultural code is maintained in the brain structures of individuals. Thus, while the behavioral systems characteristic of populations can have rules of their own, the mechanisms of change (process mechanisms) must lie rooted in the psychobiological properties of individuals. The final process of adaptation consists of encounters between systems and the selective action of the environment. The nature of both the "internal system" and the encounters between variations and the selecting field place serious constraints upon the "free will" of individual members of a social system. The process which operates in the selection of new traits is based on pressures induced by variables in the system and outside of it. I have resorted to psychobiological properties to explain the basis of certain process mechanisms in the development and maintenance of human behavioral systems. A full understanding of human behavior demands further research into these process mechanisms as well as process outcomes in relation to environmental parameters.

If the biological model of evolution is to be followed adaptation must be measured on the basis of some kind of comparative population statistic. This must not, however, draw us into a kind of crude "Social Darwinism." The model presented here is not "Social Darwinism" but merely "Darwinism." Population competition and replacement is only one feature of adaptation in the human species. As we have seen positive feedback or changes in maximization may occur as a result of internal processes and these need not be based on biological mismatches among individuals in which the strongest or most fit survive. Nor is it useful in most cases to rely on a comparison between individuals in the same population carrying a trait and those who do not. The esoteric knowledge of a few may contribute to the maximization of an entire group. An adaptive trait needs to be accepted by some but not all individuals in a population. To be adaptive it must become an accepted part of the behavioral system. We must therefore consider as did V. G. Childe, cited in Chapter VIII, the contribution of traits to the population carrying capacity of a particular environment including changes through time. We must also analyze the ability of a population to maintain itself and its behavioral system in the face of parametric shifts, i.e. how good a self-regulating system it is and what the elements which contribute to effective self-regulation are.

Population Size As a Measure of Adaptation

Population size can be used as a measure of adaptation in the study of human cultural evolution only when comparisons are made through time between populations competing for the same niche or for single populations studied through time in relation to demography, carrying capacity, and technological change. Carrying capacity itself is a difficult concept to deal with on the empirical level, although it serves a very necessary function for the construction of theory. J. Street, in "An evaluation of the concept of carrying capacity" (1969), points out that previous measures undertaken by both anthropologists and geographers are probably wide of the mark because they ignore a host of variables including the role of insect pests on cultivated fields. So far, the most fruitful studies of adaptation are those undertaken by archaeologists. As I have already pointed out, adaptation as a process can and should also be studied from the point of view of mechanisms of continuity and variation as they operate in systems of positive and negative feedback. This can be done by ethnologists as well as by archaeologists.

One problem remains to be discussed, that is, the diversion of energy in human cultural systems away from the production of organisms.

In egalitarian societies, most production goes into subsistence and that which does not is generally distributed equally. Therefore such populations operate much like other animal populations (in terms of energy conversion), and levels of adaptation can be measured accordingly. Once status differences develop (classes, castes, or external exploitation), however, energy is diverted into varying but often large quantities of nonconsumable goods which along with consumable goods are distributed unequally. In these systems much more energy is produced than is used in the production of organisms. Yet it is clear from both the anthropological and historical record that status differences in human social groups have often contributed to their further differentiation and development. The system of unequal rewards and the drive to produce new items of technology, which is the hallmark of the capitalist system, led to a tremendous outburst of innovation. The force that drives evolution in these cases cannot be strictly conceived of as population, nor can population size alone be used to compare systems of this type. The most interesting questions of social development and of evolution in the modern world therefore can only be dealt with on the level of socioeconomic analysis. Nonetheless, technological advance in nonegalitarian societies has in fact led to tremendous increases in carrying capacity and population. It appears as if the inequality of distribution has been an evolutionary cost for an effective adaptation in the Darwinian sense. It must be remembered that thoughts about population control are of recent vintage and must be understood in the context of changing economic needs as well as of the possibility that the maximal carrying capacity of our advanced technological system may soon be achieved. Advanced capitalist and socialist societies today depend more on internal consumption than they do on world trade, although trade is still an important factor among them if no longer between them and the underdeveloped countries. While technologically advanced nations require more and more of the Third World's raw materials to keep their industries running, they no longer require a vast labor supply, and their own growing local consumption has tended to replace foreign countries as the *major* source of new markets.

From the evolutionary point of view it must be remembered that these systems consume a vast quantity of resources of all types. We in the United States are all aware that raw material is being converted almost directly and at increasing rates into pollution via goods that have a very short useful life (if they have a useful life at all). Thus our productive capacities may be outstripping the net world carrying capacity. That is to say, the level at which *we* live may be much too high for the current world population. The irony of this is, of course, that the quality of life for most of the world's people is below standard from the strictly biological point of view. A return to more equal distribution at the present time with the same level of productivity and the same population would alleviate many of the world's pressing problems.

Biologically speaking, the kind of technological system that is based on capitalist enterprise but which exists in both capitalist and socialist countries may no longer be adaptive. In the past, although the full potential for energy conversion into organisms was never achieved, the mode of production and social rewards did achieve significant increases in carrying capacity, some of which were translated into organisms. The period of early capitalism was tremendously creative. Today the same system may spell doom for the human species. An analysis of the modern world from the point of view of the ecological principles presented here might be useful for a better understanding of contemporary cultural processes as they affect the future of all mankind.

A more difficult point concerns the quality of life. For lower organisms, the quality of life is controlled strictly by genetic processes that affect the level of adaptation. For human beings, the question is much more difficult. Beyond the simple fact that we all require a certain number of calories and the proper amounts of other nutritional elements per day to survive and that we all require some shelter from the elements, the fact of being human produces what we might call "felt needs" that vary in degree and kind from culture to culture. Just what level of felt needs is appropriate cannot be determined scientifically, although we certainly can scientifically investigate the probable effects that various kinds and levels of felt needs will have on the species and its environment. At the present time such questions appear to be of prime importance for all of us.

In regard to the above discussion it is necessary to point out here that the *investigation* of human biocultural evolution does not *demand* a value orientation. The aim of such research is not the discovery of the good society. We are interested in how human social systems work and how they develop, but strictly speaking as scientists, we are not interested in such value-loaded concepts as progress or happiness. A balanced system may be one in which babies die in great numbers or in which a large number of individuals are unhappy. The question of the good society is one for ethical philosophers and political scientists, not evolutionists. Nonetheless, the data provided by evolutionary studies may help the philosopher and the social planner to clarify their thinking about process so that intelligent decisions can be made about the derivation and implementation of ethical judgments. Let us hope also that ethical judgments will help evolutionists to choose worthy topics for investigation. For while scientific investigation as such must remain objective, there is no reason why the scientist should be objective in his choice of topics for study.

Summary and Conclusion

Human populations as evolutionary systems depend upon certain mechanisms of variation and continuity for their generation and maintenance. Some of these mechanisms are somatically based and are identical to those found in other animal species. Others are related to culturally based behavior which is largely extrasomatic. The somatic element in culture is restricted to the evolution of capacities for behavior, including, perhaps, some as yet unproved population-specific behavioral tendencies. In the cultural system, variation is produced through the operation of certain psychological and biological mechanisms operating on the level of the individual organism. All cultural innovations have their roots in individual behavior. Continuity in cultural systems may depend in part upon psychological and biological principles as well, but it is also maintained by requirements of the total system, which acts, as does the environment, as a selecting field on variation. The selective advantage which a new trait may carry need have no relation to the intent of individuals, although planned change certainly plays an important part in the process of human adaptation. The scientific method, for example, attempts to reduce the random nature of discovery through the development of logical theories and laws.

Both the system as a system and the environment act on the existing pool of variation to select in the direction of system maximization. The major adaptive mechanisms for man are high learning capacity and good retention, behavioral flexibility, and the code system of language and culture. Learning as an aspect of behavior provides humans with a powerful generator of variation which is constrained to some degree by the more or less orderly system which is culture. The psychobiological aspect of human behavior is linked in part to the cultural system through the development of culturally derived secondary drives and by the anxiety-reducing nature of culture. Although cultural transmission is extrasomatic, the code which produces and maintains traits is carried by individuals. However, since human groups cooperate and because a behavioral system operates on the population level, one cannot merely compare the number of individuals carrying various traits to determine the selective values of the traits. After all, not everyone need be a doctor to reduce illness in a population. Since traits affect the population as a unit, population size and measures of growth can be used as a comparative index of survival through time.

The process of human behavioral adaptation can be studied:

1. Through the testing of cross cultural hypotheses relating elements of behavior to basic biological problems.

2. By examining, through the methods of archaeology and ethnology, the adaptive value of traits in relation to outcomes between competing populations.

3. By examining the positive feedback effects of traits and trait complexes on population growth and on the carrying capacity of specific environments.

4. Through the analysis of self-regulating behavioral-populational systems in terms of their ability to withstand or respond positively to parametric change.

BIBLIOGRAPHY AND SUGGESTED READING

CHAPTER I *Darwinian Evolution and Genetics*

DARLINGTON, C. D. *The Evolution of Genetic Systems*. New York: Basic Books, 1958.

DOBZHANSKY, T. *Genetics and the Origin of Species*. New York: Columbia University Press, 1951.

HUXLEY, J. *Evolution: The Modern Synthesis*. New York: Harper, 1943.

MAYR, E. *Systematics and the Origin of Species*. New York: Dover, 1964.

—— *Animal Species and Evolution*. Cambridge: Belknap Press (Harvard), 1963.

RENSCH, B. *Evolution Above the Species Level*. London: Methuen, 1959.

SIMPSON, G. *The Major Features of Evolution*. New York: Columbia University Press, 1953.

TAX, S. *The Evolution of Life*. Chicago: University of Chicago Press, 1960.

—— *The Evolution of Man*. Chicago: University of Chicago Press, 1960.

—— *Issues in Evolution*. Chicago: University of Chicago Press, 1960.

WILLIAMS, G. C. *Adaptation and Natural Selection*. Princeton, N.J.: Princeton University Press, 1966.

CHAPTER II *Mendelian Genetics*

GARDNER, E. J. *Principles of Genetics*. New York: Wiley, 1960.

PETERS, J. A. *Classic Papers in Genetics*. Englewood
    Cliffs, N.J.: Prentice-Hall, 1959.
STERN, C. *Principles of Human Genetics*. San Francisco:
    Freeman, 1960.

CHAPTER III   *Chemical Genetics*

JUKES, T. H. *Molecules and Evolution*. New York: Colum-
    bia University Press, 1966.
LWOFF, A. *Biological Order*. Cambridge: M.I.T. Press,
    1962.
STRAUSS, B. S. *An Outline of Chemical Genetics*. Phila-
    delphia: Saunders, 1960.

CHAPTER IV   *Population Genetics and Evolution*

FALCONER, D. S. *Introduction to Quantitative Genetics*.
    New York: Ronald Press, 1960.
LI, C. C. *Human Genetics*. New York: McGraw-Hill,
    1961.
NEEL, J. V., and SCHULL, W. J. *Human Heredity*. Chi-
    cago: University of Chicago Press, 1954.
WRIGHT, S. *Evolution and the Genetics of Populations*.
    Chicago: University of Chicago Press, 1968.

CHAPTER V   *Evolutionary Systems*

ASHBY, W. R. *Design for a Brain*. New York: Wiley, 1960.
KLOPFER, P. H. *Behavioral Aspects of Ecology*. Engle-
    wood Cliffs, N.J.: Prentice-Hall, 1962.
—— *Habitats and Territories*. New York: Basic Books,
    1969.
MARGALEF, R. *Perspectives in Ecological Theory*. Chi-
    cago: University of Chicago Press, 1968.
RADCLIFFE-BROWN, A. R. N. *Structure and Function in
    Primitive Society*. Glencoe, Ill.: Free Press, 1952.

CHAPTER VI *Behavioral Genetics*

ADAMSON, JOY *Born Free.* New York: Pantheon, 1960.

BRELAND, K., and BRELAND, M. The Misbehavior of organisms. *American Psychologist,* 16:681–84, 1961.

BRUELL, J. H. Inheritance of behavioral and physiological characteristics of mice and the problem of heterosis. In *Readings in Animal Behavior,* ed. by T. E. McGill. New York: Holt, Rinehart and Winston, 1965.

FULLER, J. L., and THOMPSON, W. R. *Behavior Genetics.* New York: Wiley, 1960.

GOTTESMAN, I. I. Biogenetics of race and class. In *Social Class, Race, and Psychological Development,* ed. by M. Deutsch, I. Katz, and A. Jensen. New York: Holt, Rinehart and Winston, 1968.

HIRSCH, J. Behavior genetics and individuality understood. *Science,* 142:1436–42, 1963.

HUXLEY, J., MAYR, E., OSMOND, H., and HOFFER, A. Schizophrenia as a genetic morphism. *Nature,* 204: 220–21, 1964.

MC CLEARN, G., and RODGERS, D. A. Genetic factors in alcohol preference of laboratory mice. *Journal of Comparative and Physiological Psychology,* 54:116–19, 1961.

SCOTT, J. P., and FULLER, J. L. *Genetics and the Social Behavior of the Dog.* Chicago: University of Chicago Press, 1965.

SHELDON, W. H., STEVENS, S. S., and TUCKER, W. B. *The Varieties of Human Physique.* New York: Harper, 1940.

SHERWOOD, J. J., and NATAUPSKY, M. Predicting the conclusions of Negro-white intelligence research from biographical characteristics of the investigator. *Journal of Personality and Social Psychology,* 8, part 1:53–58, 1968.

TRYON, R. C. Genetic differences in maze-learning ability

in rats. *Yearbook of the National Society for Studies in Education,* 1940.

CHAPTER VII *Behavioral Evolution*

ALLAND, A., JR. *The Human Imperative.* New York: Columbia University Press, 1972.

ARDREY, R. *African Genesis.* New York: Atheneum, 1961.

—— *The Territorial Imperative.* New York: Atheneum, 1966.

—— *The Social Contract.* New York: Atheneum, 1970.

ARONSON, L. R., TOBACH, E., LEHRMAN, D. S., and ROSEN-BLATT, J. S., EDS. *Development and Evolution of Behavior: Essays in Memory of T. C. Schneirla.* San Francisco: Freeman, 1970.

BITTERMAN, M. E. Toward a comparative psychology of learning. *American Psychologist,* 15: 704–12, 1960.

COLD SPRINGS HARBOR SYMPOSIA ON QUANTITATIVE BIOLOGY. *Origin and Evolution of Man,* vol. 15, 1950.

DOBZHANSKY, T. *Mankind Evolving.* New Haven: Yale University Press, 1962.

ETKIN, W., ED. *Social Behavior and Organization Among Vertebrates.* Chicago: University of Chicago Press, 1964.

JAY, P. C., ED. *Primates: Studies in Adaptation and Variability.* New York: Holt, Rinehart and Winston, 1968.

JENNINGS, H. S. *Behavior of the Lower Organisms.* Bloomington, Ind.: University of Indiana Press, 1958.

JOLLY, A. *Lemur Behavior.* Chicago: University of Chicago Press, 1966.

—— *The Evolution of Primate Behavior.* New York: Macmillan, 1972.

JOLLY, C. J. The seed eaters: a new model of hominid differentiation based on a baboon analogy. *Man,* n.s., 5:5–26, 1970.

LORENZ, K. *On Aggression.* New York: Harcourt, Brace & World, 1966.

—— *Studies in Animal and Human Behavior*. Cambridge: Harvard University Press, 1970–72.

MORRIS, D. *The Naked Ape*. New York: McGraw-Hill, 1967.

ROE, A., and SIMPSON, G. G. *Behavior and Evolution*. New Haven: Yale University Press, 1958.

SCHALLER, GEORGE. *The Mountain Gorilla: Ecology and Behavior*. Chicago: University of Chicago Press, 1963.

—— *The Year of the Gorilla*. Chicago: University of Chicago Press, 1964.

SCHILLER, C. H., and LASHLEY, K. S. *Instinctive Behavior*. New York: International Universities Press, 1957.

SCOTT, J. P. *Animal Behavior*. Chicago: University of Chicago Press, 1958.

SPUHLER, J. N. *The Evolution of Man's Capacity for Culture*. Detroit: Wayne State University Press, 1959.

—— Somatic Paths to Culture. In *The Evolution of Man's Capacity for Culture*, ed. by J. N. Spuhler. Detroit: Wayne State University Press, 1959.

TIGER, LIONEL. *The Imperial Animal*. New York: Holt, 1971.

WASHBURN, S. L. Tools and human evolution. In *Human Variations and Origins*, ed. by W. S. Laughlin and R. H. Osborne. San Francisco: Freeman, 1967.

CHAPTER VIII *Culture and Human Behavior*

ADAMS, R. M. *The Rise of Urban Society*. Chicago: University of Chicago Press, 1969.

ALLAND, A., JR. *Adaptation in Cultural Evolution: An Approach to Medical Anthropology*. New York: Columbia University Press, 1970.

BARTH, F. Ecological relationships of ethnic groups in Swat, North Pakistan. *American Anthropologist*, 58: 1079–89, 1956.

BROWN, R. *Words and Things*. Glencoe, Ill.: Free Press, 1958.

CARNEIRO, R. Scale analysis as an instrument for the study of cultural evolution. *Southwestern Journal of Anthropology*, 18:149–69, 1962.

CHILDE, V. G. *Man Makes Himself*. London: Watts, 1936.

—— *Social Evolution*. New York: Schuman, 1952.

COE, M., and FLANNERY, K. V. Micro environments and Mesoamerican prehistory. In *New Roads to Yesterday*, ed. by J. Caldwell. New York: Basic Books, 1966.

DOLE, G. E. The classification of Yankee nomenclature in the light of evolution in kinship. In *Essays in the Science of Culture in Honor of Leslie White*, ed. by G. E. Dole and R. Carneiro. New York: Crowell, 1960.

FLANNERY, K. V. The ecology of early food production in Mesopotamia. In *Environment and Cultural Behavior*, ed. by A. P. Vayda. Garden City, N.Y.: Natural History Press, 1969.

—— Archeological systems theory and early Mesoamerica. In *Anthropological Archeology in the Americas*, ed. by B. J. Meggers. The Anthropological Society of Washington, D.C., 1968.

FRIED, M. *The Evolution of Political Society*. New York: Random House, 1968.

MALINOWSKI, B. *A Scientific Theory of Culture*. Chapel Hill: University of North Carolina Press, 1944.

MORGAN, L. H. *Ancient Society*. New York: Holt, 1877.

MURPHY, R. F., and STEWARD, J. Tappers and trappers: parallel process in acculturation. *Economic Development and Cultural Change*, 4:335–53, 1956.

OSGOOD, C. Ingalik mental culture. *Yale University Publications in Anthropology*, No. 56, 1959.

RADCLIFFE-BROWN, A. R. N. *Structure and Function in Primitive Society*. Glencoe, Ill.: Free Press, 1952.

RAPPAPORT, R. A. *Pigs for the Ancestors*. New Haven: Yale University Press, 1968.

SAHLINS, M. D. Notes on the original affluent society. In *Man the Hunter,* ed. by R. Lee and I. DeVore. Chicago: Aldine, 1968.

——, and SERVICE, E. R. *Evolution and Culture.* Ann Arbor: University of Michigan Press, 1960.

SPENCER, H. *First Principles.* New York: Appleton, 1864.

STEWARD, J. *Theory of Culture Change.* Urbana, Ill.: University of Illinois Press, 1955.

TYLOR, E. B. *Anthropology.* New York: Appleton, 1928 (first pub. 1896).

—— *Primitive Culture.* London: John Murray, 1871.

VAYDA, A. P., and RAPPAPORT, R. Ecology, cultural and noncultural. In *Introduction to Cultural Anthropology,* ed. by James Clifton. Boston: Houghton Mifflin, 1966.

WHITE, L. *The Evolution of Culture.* New York: McGraw-Hill, 1959.

CHAPTER IX *Culture and Evolution*

ASHBY, W. R. *Design for a Brain.* New York: Wiley, 1960.

BARNETT, H. *Innovation, the Basis of Culture Change.* New York: McGraw-Hill, 1953.

KROEBER, A. L. *Cultural and Natural Areas of Native North America.* Berkeley: University of California Press, 1939.

LÉVI-STRAUSS, C. *Les Mythologiques,* 4 vols. Paris: Plon, 1964–72.

MEAD, M. *Continuities in Cultural Evolution.* New Haven: Yale University Press, 1964.

MURDOCK, G. P. *Social Structure.* New York: Macmillan, 1949.

—— *Africa: Its People and Their Cultural History.* New York: McGraw-Hill, 1959.

SAHLINS, M. D., and SERVICE, E. R. *Evolution and Culture.* Ann Arbor: University of Michigan Press, 1960.

SCOTT, J. P., and FULLER, J. L. *Genetics and the Social*

Behavior of the Dog. Chicago: University of Chicago Press, 1965.

SIMPSON, G. *The Major Features of Evolution.* New York: Columbia University Press, 1953.

CHAPTER X   *The Adaptive Model*

ANTHROPOLOGICAL SOCIETY OF WASHINGTON. *Evolution and Anthropology: A Centennial Appraisal,* 1959.

BOSERUP, E. *The Conditions of Agricultural Growth.* Chicago: Aldine, 1965.

BRUNER, E. Primary group experience and the process of acculturation. *American Anthropologist,* 58:605–23, 1956.

GOODMAN, M. Man's place in the phylogeny of the primates as reflected in serum proteins. In *Classification and Human Evolution,* ed. by Sherwood L. Washburn. Chicago: Aldine, 1963.

HARNER, M. J. Population pressure and the social evolution of agriculturists. *Southwestern Journal of Anthropology,* 26:67–86, 1970.

HARRIS, M. The cultural ecology of India's sacred cow. *Current Anthropology,* 7:51–66, 1966.

MOORE, O. K., and ANDERSON, A. R. The structure of personality. *The Review of Metaphysics,* 16:212–36, 1962.

STREET, J. An evaluation of the concept of carrying capacity. *Professional Geographer,* 21:104–7, 1969.

RAPPAPORT, R. A. *Pigs for the Ancestors.* New Haven: Yale University Press, 1968.

WHITING, J. Effects of climate on certain cultural practices. In *Explorations in Cultural Anthropology: Essays in Honor of G. P. Murdock,* ed. by Ward Goodenough. New York: McGraw-Hill, 1964.

WILLIAMS, G. R., FISCHER, A., FISCHER, J. L., and KURLAND, L. T. An evaluation of the kuru hypothesis. *Journal Génétique Humaine,* 13:11–21, 1964.

INDEX